Collins

Grade A/A* Booster Workbook

Get your Grade A*

NEW GCSE MATHS
Edexcel Linear
Matches the 2010 GCSE Specification

Greg Byrd

William Collins' dream of knowledge for all began with the publication of his first book in 1819. A self-educated mill worker, he not only enriched millions of lives, but also founded a flourishing publishing house. Today, staying true to this spirit, Collins books are packed with inspiration, innovation and practical expertise. They place you at the centre of a world of possibility and give you exactly what you need to explore it.

Collins. Freedom to teach.

Published by Collins
An imprint of HarperCollinsPublishers
77–85 Fulham Palace Road
Hammersmith
London
W6 8JB

Browse the complete Collins catalogue at:
www.collinseducation.com

10 9 8 7 6 5 4

ISBN-13 978-0-00-741003-3

British Library Cataloguing in Publication Data
A Catalogue record for this publication is available from the British Library.

Commissioned by Katie Sergeant
Project managed by Rebecca A.S. Richardson
Grade progression maps by Claire Powis, Maths AST; Children, Families and Education, Kent
How to interpret the language of exams and Further Hints by Chris Curtis, Head of Maths, Frome Community College, Somerset
Edited and proofread by Joan Miller and Gudrun Kaiser
Answers checked by Joan Miller and Steven Matchett
Exam board specification checked by Lindsey Charles
Cover design by Angela English
Concept design by Lesley Gray
Illustrations by Kathy Baxendale
Design and typesetting by Ken Vail Graphic Design
Production by Arjen Jansen
Printed and bound in China
With thanks to Andy Manley: Education Consultant, Central Bedfordshire 'Reach for the Stars' Network.

Contents

Introduction

This workbook aims to help you reach your full potential and get an A* in your maths GCSE. It gives you plenty of practice with brand new content in the key topics and main sections of your course.

These sections are colour coded: Number, Algebra, Geometry and Measures, Statistics and Probability.

Question grades

You can tell the grade of each question or question part by the colour of its number:

Grade B questions are shown as **1**

Grade A questions are shown as **1**

Grade A* questions are shown as **1**

Hint boxes and Further hints

The 'Hint' boxes provide you with guidance as to how to approach challenging questions, and the 'Further hints' on pages 122–123 provide you with facts you should know in order to succeed in your exams.

> **Hint:** Keep thinking – square numbers.

Assessing understanding and problem solving

This section encourages assessing understanding and problem solving with a focus on the new aspects of the curriculum. It helps you to build and use your process skills and prepare for your exams.

Spot the errors

This part of the book promotes analytical thinking and problem solving. In each of the questions there are one or more incorrect student answers. You need to look at the incorrect responses in the blue and yellow boxes marked with ✎, show where the students went wrong and then work out the correct answer yourself.

Grade progression maps

On pages 101–108, you will find grade progression maps for the main subject areas. These show how you can progress from grade B to grade A and from grade A to A*. Use the progression maps to check what you know and to see how you can move up a grade.

Answers

You will find answers to all the questions at the back of the book. Check your answers yourself, or your teacher might tear them out and give them to you later, to mark your work.

How to interpret the language of exams

This section provides a helpful table that explains the wording used in examinations and how to interpret it correctly to get full marks.

Formulae sheet

Finally, you are provided with a formulae sheet outlining the standard formulae given to you in your exam so you know what to expect on the day.

Collins provides extra practice and differentiation at all levels. Remember that in your Higher exam, about 50% of your marks come from questions at the lower grades. For more practice at these grades, see these books:

Workbooks – focus on grades G, F, E

Grade C Booster Workbooks – focus on grades D, C, B

Grade A/A* Booster Workbooks – focus on grades B, A, A*

Estimating with square roots

Hint: Keep thinking – square numbers.

1 Give the answer to each of the following.

a $\sqrt{0.04}$ = _____

b $\sqrt{0.25}$ = _____

c $\sqrt{0.64}$ = _____

d $\sqrt{1.21}$ = _____

e $\sqrt{2.25}$ = _____

f $\sqrt{0.0144}$ = _____

g $\sqrt{0.000169}$ = _____

h $\sqrt{0.000081}$ = _____

i $\sqrt{0.0025}$ = _____

2 Estimate the value of each of the following.

a $\sqrt{34.5}$ = _____

b $\sqrt{219.8}$ = _____

c $\sqrt{1.36}$ = _____

d $\sqrt{0.088}$ = _____

e $\sqrt{0.38}$ = _____

f $\sqrt{388.75}$ = _____

3 Estimate the value of each of the following. Show your working.

a $18.7 \times \sqrt{10.01}$ _____

b $5 + \sqrt{0.77}$ _____

c $\sqrt{\dfrac{29.44^2}{8.75}}$ _____

d $\dfrac{\sqrt{4.88 \times 5.45}}{0.472}$ _____

e $\sqrt{0.22} \times \sqrt{1.5}$ _____

f $\dfrac{2.8^2 + 6.155^2}{\sqrt{82.56}}$ _____

g $\sqrt{(\sqrt{2.11} - \sqrt{0.8})}$ _____

h $\sqrt{1.2} + \sqrt{1.23} + \sqrt{123.4}$ _____

i $\dfrac{10 - \sqrt{9.09}}{\sqrt{0.5}}$ _____

☑ Reverse percentages

1 A Formula One racing car tyre loses approximately 10% of its weight during a race, due to wear. After a race a complete set of tyres weighs about 18 kg. What do they weigh before the race?

Hint: Don't just find 10% and add it on!

_____

_____

2 A Formula One racing car driver loses approximately 5% of his or her body weight during a race. Lewis Hamilton weighed 64.6 kg after a race. How much did he weigh before the race?

_____

_____

3 A set of tyres for a Formula One racing car costs around £3000 including VAT at 17.5%. How much does a set of tyres cost before the VAT is added? Give your answer to the nearest £10.

_____

_____

4 In 2009, after a 150% increase in salary, Jenson Button was estimated to earn around 18 million dollars. How much did he earn in 2008?

_____

_____

5 During a race the disc brakes of a Formula One racing car heat to 1000 °C. This causes them to expand by 0.15%. At maximum expansion, a disc has a diameter of 278.417 mm. What is the normal diameter of the disc brake?

_____

_____

6 If, two years ago, David Coulthard had put all his Formula One earnings into a bank account and earned 5% compound interest, he would now have $7 938 000 in his account. How much did he earn two years ago?

_____

_____

✅ Standard index form

1 The Avogadro constant is 602 300 000 000 000 000 000 000.

Express this in standard index form. _____

2 The total volume of seawater on Earth is about 1 370 000 000 000 000 000 m^3.

Express this in standard index form. _____

3 A single cold virus is 0.000 000 02 metres long.

Express this in standard index form. _____

4 One atom of gold has a mass of 0.000 000 000 000 000 000 000 000 33 grams.

Express this in standard index form. _____

5 Write each of the following as ordinary numbers.

a 1×10^4 _____   **b** 1.2×10^5 _____

c 1.23×10^{-5} _____   **d** 1.234×10^{-1} _____

6 Work out each of the following. Give all your answers in standard index form.

a $(3 \times 10^4) + (4 \times 10^3)$ _____

b $(3 \times 10^4) - (4 \times 10^3)$ _____

c $(3 \times 10^4) \times (4 \times 10^3)$ _____

d $(3 \times 10^4) \div (4 \times 10^3)$ _____

7 The speed of light is 300 000 km/s. Light takes $8\frac{1}{2}$ minutes to travel from the Sun to the Earth. How far is the Sun from the Earth, in centimetres? Express this distance in standard index form.

_____

8 You probably have around 2×10^{13} red corpuscles in your bloodstream. Each red corpuscle weighs about 0.000 000 000 1 grams. Work out the total mass of your red corpuscles, in kilograms. Give your answer in standard index form.

_____

9 A factory produces 3.6×10^6 nails each year. Each nail has a mass of 5×10^{-3} kg. Of all nails produced, 0.75% are faulty. Work out the total mass of all the faulty nails produced in one year.

_____

✅ Limits

1 Write down the range of possible values of each of the following.

a 5.7 kg (to 1 dp) _____

b 5.20 mg (to 3 sf) _____

c 89.999 Gb (to 3 dp) _____

2 Tamara loads 20 boxes of machine parts onto a wooden pallet. Each box weighs 45 kg, to the nearest kilogram, and the wooden pallet weighs 18 kg, to the nearest kilogram.

What is the maximum weight of the pallet and 20 boxes?

_____

_____

3 A cube has a side length of 10 cm, measured to the nearest centimetre.

a What is the minimum surface area of the cube?

_____

b What is the maximum volume of the cube?

_____

_____

4 A card measuring 12.5 cm by 8.5 cm (both measured to the nearest 0.1 cm) is to be posted in an envelope that is 13 cm by 10 cm (both to the nearest 1 cm).

Can you guarantee that the card will fit the envelope? Explain your answer.

_____

_____

5 Sammi runs a 50 m race at a speed of 6 m/s.
Both values are measured to an accuracy of 1 significant figure.

a What is Sammi's fastest possible time?

_____

She has an average stride length of 1.32 m, to the nearest centimetre.

b What is the smallest number of steps she will take?

_____

Indices

1 Write down the value of each of the following.

a 1^0 _____ **b** 5^0 _____ **c** n^0 _____

2 Write down each of these in fraction form.

a 2^{-3} _____ **b** 5^{-3} _____ **c** 10^{-4} _____

d 12^{-1} _____ **e** x^{-1} _____ **f** x^{-n} _____

3 Write down each of these in negative index form.

a $\dfrac{1}{3^2}$ _____ **b** $\dfrac{1}{t^3}$ _____ **c** $\dfrac{1}{h^m}$ _____

4 Evaluate the following.

a $4^4 \times 4^{-2}$ _____ **b** $4^2 \div 4^{-2}$ _____

c $8^4 \times 8^{-4}$ _____ **d** $\dfrac{2^5}{2^9}$ _____

5 Write down each of these in fraction form.

a $2t^{-3}$ _____ **b** $\dfrac{3}{4}p^{-1}$ _____

6 Evaluate each of the following.

a $25^{\frac{1}{2}}$ _____ **b** $125^{\frac{1}{3}}$ _____

c $64^{-\frac{1}{3}}$ _____ **d** $16^{\frac{1}{4}}$ _____

e $100^{0.5}$ _____ **f** $32^{-\frac{1}{5}}$ _____

7 Evaluate each of the following.

a $8^{\frac{2}{3}}$ _____ **b** $64^{\frac{2}{3}}$ _____

c $1000^{\frac{2}{3}}$ _____ **d** $4^{\frac{5}{2}}$ _____

e $27^{-\frac{2}{3}}$ _____ **f** $100^{-\frac{5}{2}}$ _____

Recurring decimals to fractions

1 Convert each of these recurring decimals to a fraction in its simplest form.

a $0.\dot{1}$ _____

b $0.\dot{2}$ _____

c $0.\dot{3}$ _____

d $0.\dot{4}$ _____

e $0.\dot{7}$ _____

f $0.\dot{8}$ _____

2 Ben says that $0.\dot{9}$ is equivalent to 1. Claire thinks he is wrong.

How can Ben use recurring decimals and fractions to show Claire that $0.\dot{9}$ is equivalent to 1?

_____

_____

_____

3 Convert each of these recurring decimals to a fraction in its simplest form.

a $0.\dot{2}\dot{3}$ _____

b $0.0\dot{2}$ _____

c $0.24\dot{3}$ _____

d $0.2\dot{2}0\dot{4}$ _____

Surds

Hint: Remember to find the square root of any square number.

1 Write each of these expressions as a single square root.

   a $\sqrt{3} \times \sqrt{5}$ _____

   b $\sqrt{3} \times \sqrt{2} \times \sqrt{10}$ _____

2 Work out the value of each of the following.

   a $\sqrt{5} \times \sqrt{5}$ _____

   b $\sqrt{3} \times \sqrt{2} \times \sqrt{6}$ _____

   c $\sqrt{10} \times \sqrt{40}$ _____

   d $\sqrt{6} \times \sqrt{2} \times \sqrt{12}$ _____

   e $\sqrt{600} \div \sqrt{6}$ _____

   f $\sqrt{63} \div \sqrt{7}$ _____

3 Write out each of these expressions in the form $a\sqrt{b}$ where b is a prime number.

   a $\sqrt{12}$ _____

   b $\sqrt{80}$ _____

4 Simplify these expressions. Write your answers as surds where necessary.

   a $3\sqrt{5} \times 2\sqrt{3}$ _____

   b $3\sqrt{8} \times 3\sqrt{3}$ _____

   c $\dfrac{4\sqrt{30}}{\sqrt{6}}$ _____

   d $\dfrac{8\sqrt{125}}{2\sqrt{20}}$ _____

   e $\sqrt{50} + 2\sqrt{32}$ _____

   f $6\sqrt{12} - 3\sqrt{27}$ _____

5 Evaluate each of the following.

   a $\dfrac{18}{\sqrt{5}} \times \dfrac{\sqrt{20}}{3}$ _____

   b $\dfrac{15\sqrt{70}}{\sqrt{5}} \times \dfrac{2\sqrt{2}}{3\sqrt{7}}$ _____

Surds

6 Simplify each of the following by rationalising the denominator.

a $\dfrac{12}{\sqrt{3}}$ _____

b $\dfrac{12}{\sqrt{8}}$ _____

c $\dfrac{3\sqrt{5}}{2\sqrt{45}}$ _____

d $\dfrac{4}{1+\sqrt{3}}$ _____

7 Expand the brackets and simplify.

a $(2+\sqrt{3})(4+\sqrt{3})$ _____

b $(4-3\sqrt{3})(5+4\sqrt{3})$ _____

c $(1-2\sqrt{7})^2$ _____

8 Show that $(\sqrt{15}-\sqrt{12})(\sqrt{15}+\sqrt{12})=3$

_____

_____

9 Simplify $\dfrac{8+\sqrt{27}}{4}-\dfrac{2+2\sqrt{3}}{3}$

_____

_____

10 **a** Show that this triangle is right-angled.

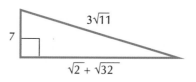

_____

_____

_____

b Find the exact area of the triangle.

_____

_____

Solving quadratic equations graphically

1 The graph on the right shows the curve $y = x^2 + 2x - 3$

> **Hint:** The roots of a quadratic equation are the values of x for certain values of y, often zero (0).

a Write down the roots of the equation $x^2 + 2x - 3 = 0$

_____

b By drawing the line $y = 1$ on the graph, solve the equation $x^2 + 2x - 4 = 0$

_____

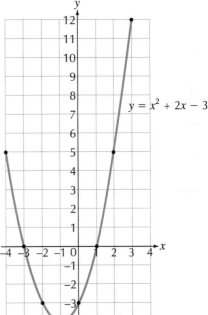

$y = x^2 + 2x - 3$

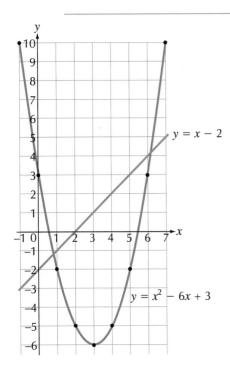

$y = x - 2$

$y = x^2 - 6x + 3$

2 The graph on the left shows $y = x^2 - 6x + 3$ and $y = x - 2$

a What is the quadratic equation for which the solutions are the x-coordinates of the points of intersection of $y = x^2 - 6x + 3$ and $y = x - 2$?

> **Hint:** Put $x^2 - 6x + 3 = x - 2$, then rearrange.

_____

b What are the solutions to the equation formed in part **a**?

_____

3 The graph on the right shows the curve $y = x^2 - 3x + 2$

By drawing a suitable straight line, solve the equation $x^2 - 4x + 1 = 0$

_____

_____

> **Hint:** Rearrange $x^2 - 3x + 2$ to get $x^2 - 4x + 1$

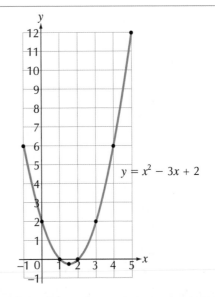

$y = x^2 - 3x + 2$

Recognising shapes of graphs

Hint: Match the obvious ones first.

Match each graph with its equation.

1 $y = x^2 + 4$ is graph _____

2 $y = 2x^2 + 4$ is graph _____

3 $y = x^2 - 4$ is graph _____

4 $y = x^2 + 2x$ is graph _____

5 $y = 2x + 4$ is graph _____

6 $y = x^3 + 4$ is graph _____

7 $y = -x^3 + 4$ is graph _____

8 $y = x^2 + 2x + 4$ is graph _____

9 $y = \dfrac{4}{x}$ is graph _____

A

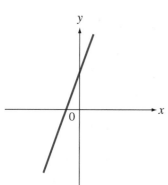

B

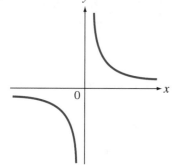

C

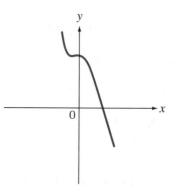

D

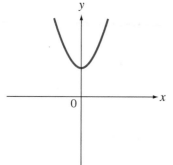

E

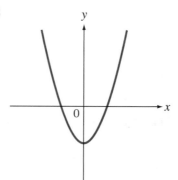

F

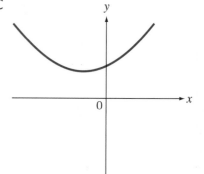

G

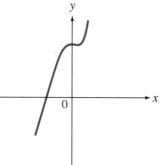

H

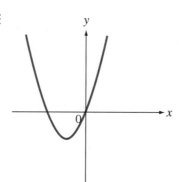

I

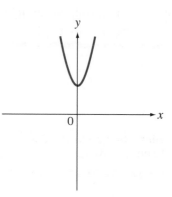

ALGEBRA

✔ Real-life graphs

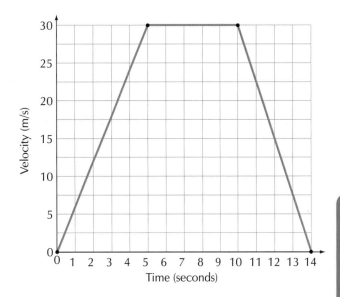

1 The graph shows the velocity–time graph for a car journey.

Work out:

a the acceleration in the first 5 seconds

_____

b the deceleration in the last second

_____

c the total distance travelled

_____

_____

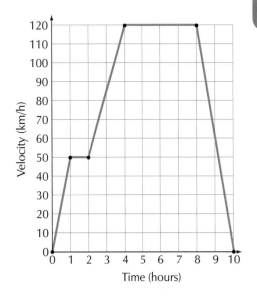

2 The graph shows the velocity–time graph for a train journey.

Work out:

a the acceleration in the first minute

_____

b the deceleration in the last minute

_____

c the acceleration between $t = 2$ and $t = 4$ hours

_____

d the distance travelled in the first two hours

_____

_____

e the distance travelled in the last two hours

_____

_____

Real-life graphs

3 15 years ago Abby invested £1000 in an account that paid 9% compound interest per year.

The interest is added at the end of each year.

Abby neither added nor withdrew any money from the account.

Hint: Remember that compound interest is different from simple interest.

a Draw a graph of $A = 1000 \times 1.09^y$ for $0 \leqslant y \leqslant 15$

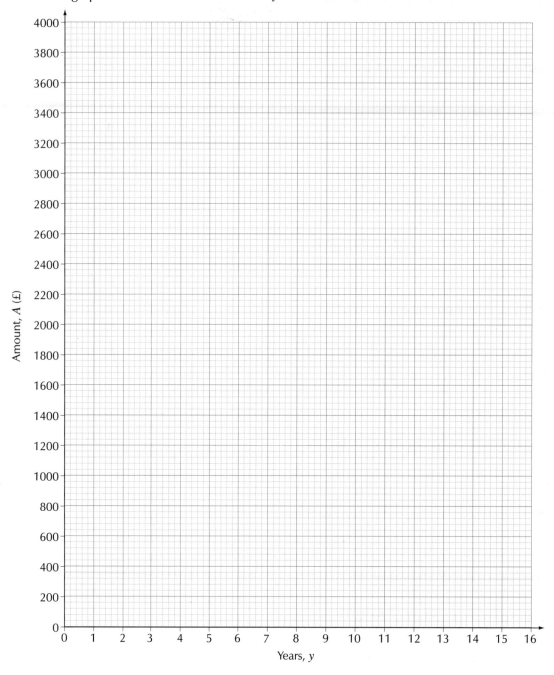

b Use your graph to find the number of years it took for the initial amount to triple.

✓ Equation of a straight line: $y = mx + c$

Hint: Don't draw the lines, just use the x- and y-values from the coordinates.

1 Work out the gradient of the line joining these points.

a (2, 1) and (4, 9)

$$\frac{9-1}{4-2} = \frac{8}{2} = 4$$

$$y = 4x + c$$

b (−3, 1) and (2, −3)

$$\frac{-3-1}{2--3} = \frac{-4}{1}$$

2 Work out the equation of the line joining these points.

a (2, 5) and (6, 17)

b (2, 1) and (6, −7)

3 **a** Work out the equation of the line with a gradient of 2 that passes through (2, 3).

b Work out the equation of the line with a gradient of $\frac{1}{2}$ that passes through (6, 0).

4 What is the equation of the line shown?

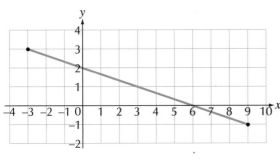

5 Work out the equation of the red line.

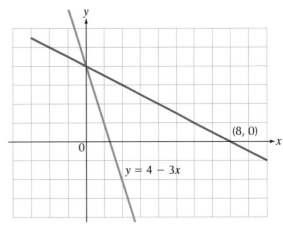

$y = 4 - 3x$

(8, 0)

ALGEBRA

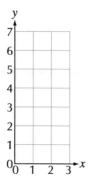 Solving simultaneous equations graphically

Hint: Find three coordinate points for each line.

Solve each pair of simultaneous equations graphically.

1 $y = -x + 7$

$y = 3x + 3$

$x = $ _____ $y = $ _____

2 $2x + 6y = 26$

$5x + 2y = 13$

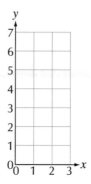

$x = $ _____ $y = $ _____

3 $y = x^2 - 2x$

$y = x + 4$

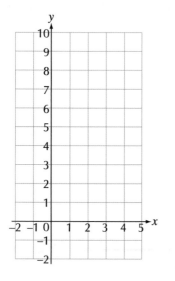

$x = $ _____ $y = $ _____

$x = $ _____ $y = $ _____

4 $y = 2x^2$

$y = 2x + 4$

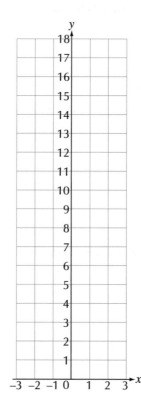

$x = $ _____

$y = $ _____

$x = $ _____

$y = $ _____

ALGEBRA

Inequalities

1 Use inequalities to describe the shaded regions.

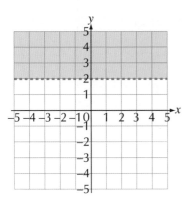

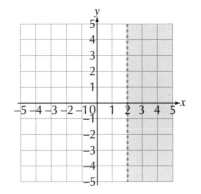

a _____

b _____

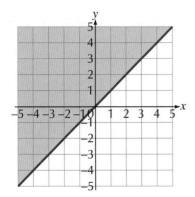

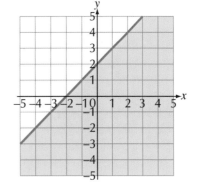

c _____

d _____

2 Draw a graph to show the region that satisfies each inequality.

a $x \geqslant -1$ **b** $y > 1$ **c** $y > x - 1$

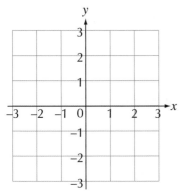

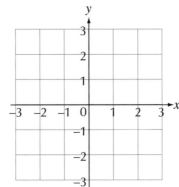

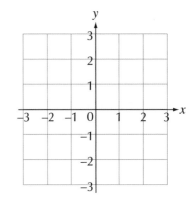

Inequalities

3 Write down the three inequalities that together describe the shaded region.

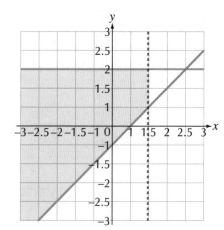

_____

_____

_____

4 Draw graphs to show the region defined by these three inequalities.

$x > -2$ $y \leq 1$ $y < -2x + 3$

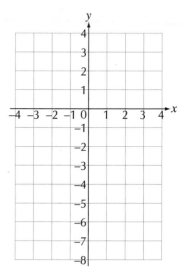

5 **a** Using the grid below, draw the graphs of $xy = 8$ and $x + y = 8$

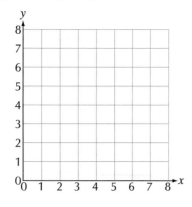

b Find all pairs of positive integers with product greater than 8 and sum less than 8.

_____

_____

Drawing complex graphs

1 **a** Complete the table of values for the function $y = x^3 + 2x^2 - 1$

x	−2.5	−2	−1	0	1	1.2
x^3		−8		0	1	
$+2x^2$		8		0	2	
-1		−1		−1	−1	
$y = x^3 + 2x^2 - 1$		−1		−1	2	

b Draw the graph of $y = x^3 + 2x^2 - 1$ for $-2.5 \leqslant x \leqslant 1.2$

Hint: The graph should be a smooth curve – don't use a ruler.

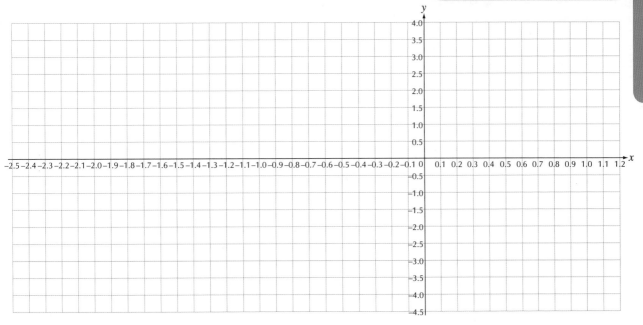

c Find the solutions to $x^3 + 2x^2 - 1 = 0$

_____

_____

_____

_____

_____

Drawing complex graphs

2 **a** Complete the table of values for the function $f(x) = \dfrac{12}{x} + 2$

x	−6	−5	−4	−3	−2	−1		1	2	3	4	5	6
$f(x)$	0	−0.4			−4			14			5		

b Draw the graph of $f(x) = \dfrac{12}{x} + 2$ for $-6 \leqslant x \leqslant 6$

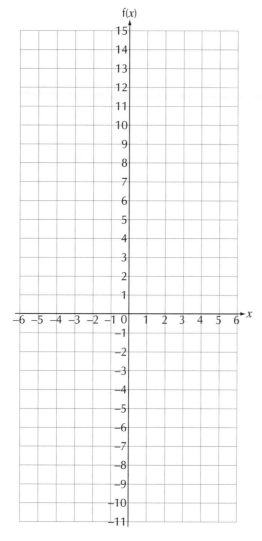

c Name the equations of the two asymptotes of $f(x) = \dfrac{12}{x} + 2$

_____

d Use your graph to estimate the value of f(3.2)

_____

ALGEBRA

3 Bacteria on a piece of warm, raw chicken can increase at the rate of

$$N = 200 \times 2^t$$

where N is the number of bacteria and t is the number of 20-minute periods of time.

a Complete the table of values for the function $N = 200 \times 2^t$

t	0	1	2	3	4	5	6	8	10	12
N	200	400	800							819 200

b Draw the graph of $N = 200 \times 2^t$ for $0 \leqslant t \leqslant 12$

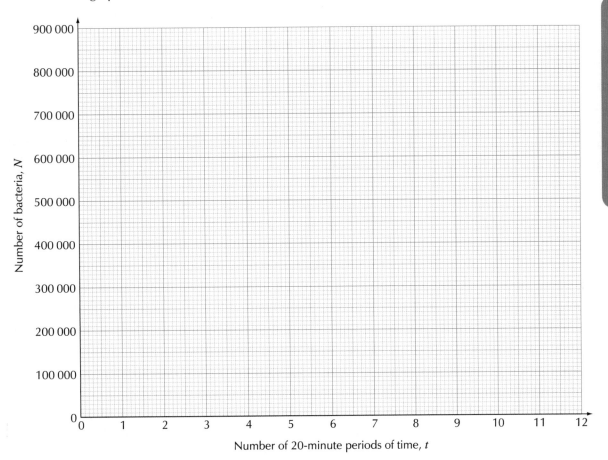

c Use the graph of $N = 200 \times 2^t$ to estimate the number of bacteria after:

i 2 hours 10 minutes _____

ii 3 hours 10 minutes _____

d Use the graph of $N = 200 \times 2^t$ to estimate the time taken for the bacteria to reach:

i 100 000 _____

ii 300 000 _____

Parallel and perpendicular graphs

1 Without plotting the straight lines, identify which lines are parallel to $y = 2x - 2$

a $y = 2x + 2$ *Yes / No*   **b** $y = -2x - 2$ *Yes / No*   **c** $y = 2 - 2x$ *Yes / No*

d $y = 2x - \frac{1}{2}$ *Yes / No*   **e** $y = \frac{1}{2}x + 2$ *Yes / No*   **f** $\frac{1}{2}y = x + 2$ *Yes / No*

2 Without plotting, identify which lines are perpendicular to $y = 2x - 2$

a $y = \frac{1}{2}x + 2$ *Yes / No*   **b** $y = -2x - 2$ *Yes / No*   **c** $y = 2 - \frac{1}{2}x$ *Yes / No*

d $y = -\frac{1}{2}x - \frac{1}{2}$ *Yes / No*   **e** $y = -2x + 2$ *Yes / No*   **f** $-\frac{1}{2}y = x + 2$ *Yes / No*

3 **a** Write down an equation of a line that is parallel to $y = -3x + 2$

b Write down an equation of a line that is perpendicular to $y = -3x + 2$

4 **a** Write down the equation of a line that is parallel to $y = -3x + 2$ and passes through the coordinate point $(-1, 9)$.

b Write down the equation of a line that is perpendicular to $y = -3x + 2$ and passes through the coordinate point $(-1, 9)$.

5 Find the equation of the perpendicular bisector of the line segment joining the points A$(-3, -2)$ and B$(7, 18)$.

Trigonometric graphs

The graphs below shows values of sin x from 0° and 360° and cos x from 0° and 360°.

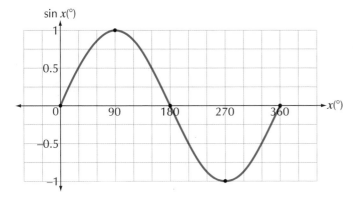

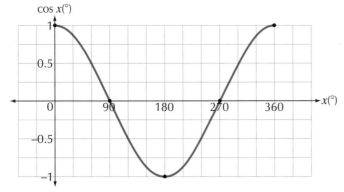

Hint: You are told to use the graphs but you will also need to use a calculator.

ALGEBRA

1 Use the graphs above to help you find the solutions to the following equations. Give all answers in the range 0° to 360° and to the nearest degree.

a sin x = 0.5878 _____

b sin x = −0.5878 _____

c cos x = 0.5878 _____

d cos x = −0.5878 _____

e sin x = 0.9993 _____

f cos x = 0.0348 _____

2 You are given that sin 29° = 0.4848.

a Find another value of x in the interval 0° ≤ x ≤ 360° for which sin x = 0.4848. _____

b Find all values of x in this interval that satisfy the equation sin x = −0.4848. _____

c Find all the solutions of the equation cos x = 0.4848 in this interval. _____

Transformation of functions

1 The graph of $y = f(x)$ is transformed.

Match the transformed function with its graph A, B or C.

a $y = f(x) + 1 =$ _____

b $y = f(x) - 1 =$ _____

c $y = f(x) - 2 =$ _____

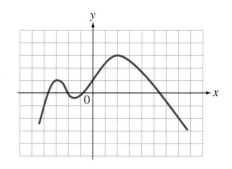

A

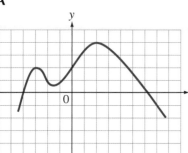

B

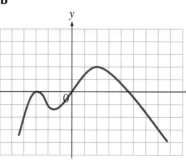

C

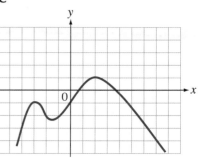

2 The graph shows the function $y = f(x)$.

Where would points A, B and C be translated to, under these transformations?

a $f(x) - 5$:

A(_____ , _____), B(_____ , _____), C(_____ , _____)

b $2 + f(x)$:

A(_____ , _____), B(_____ , _____), C(_____ , _____)

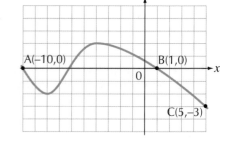

3 Sketch the graph of $y = f(x)$ for each of the given transformations.

a $f(x - 2)$

b $f(x + 2)$

c $f(1 + x)$

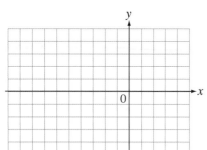

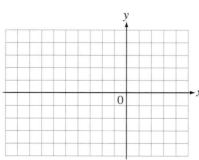

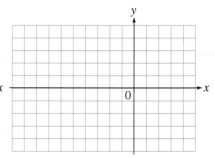

ALGEBRA

4 The graph shows the function $y = f(x)$.

Where would points A, B and C be translated to, under these transformations?

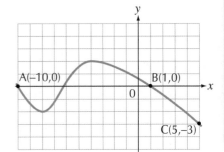

a $2f(x)$: A(_____ , _____), B(_____ , _____), C(_____ , _____)

b $-3f(x)$: A(_____ , _____), B(_____ , _____), C(_____ , _____)

c $\frac{1}{2}f(x)$: A(_____ , _____), B(_____ , _____), C(_____ , _____)

d $2f(x - 2)$: A(_____ , _____), B(_____ , _____), C(_____ , _____)

e $2f(x) + 1$: A(_____ , _____), B(_____ , _____), C(_____ , _____)

5 The graph shows the function $y = f(x)$.

Where would points A, B and C be translated to, under these transformations?

a $f(2x)$: A(_____ , _____), B(_____ , _____), C(_____ , _____)

b $f(-x)$: A(_____ , _____), B(_____ , _____), C(_____ , _____)

c $f(\frac{1}{2}x)$: A(_____ , _____), B(_____ , _____), C(_____ , _____)

6 Here is the graph of $y = \cos x$.

Use different colours to draw these graphs.

a $y = \cos x + 1$

b $y = \cos 2x$

c $y = -\cos x$

d $y = 2\cos x$

e $y = 1 - \cos (x - 90°)$

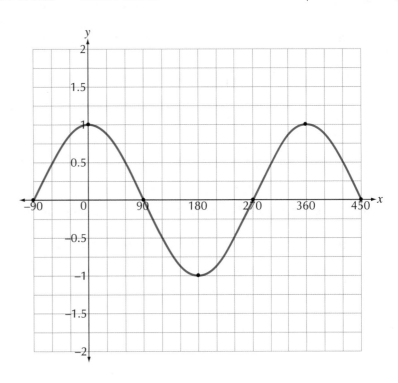

Changing the subject of a formula

1 Make r the subject of each of the following formulae.

Hint: Make sure you show each step of your working.

a $t = \sqrt{r + p}$

b $f = \sqrt{3r - e}$

c $d = 5\sqrt{r} + w$

d $V = \frac{1}{3}\pi r^2 h$

e $A = \pi(r^2 - s^2)$

f $T = 2\pi\sqrt{\dfrac{r}{g}}$

2 Make x the subject of each of the following formulae.

a $tx + y = 7 - tx$

b $4(x - y) = 2x + 3$

c $a(x + r) = b(x + t)$

3 Make x the subject of each of the following formulae.

a $\sqrt{\dfrac{3x + a}{x}} = b$

b $ax^2 + b = t - cx^2$

c $\dfrac{a}{b} = \dfrac{x}{c(d - x)}$

ALGEBRA

Solving simultaneous equations: both linear

1 Solve each pair of simultaneous equations.

   a $2x - 3y = 3$

      $2x + 5y = 27$

   b $3x + 4y = 17$

      $x + 4y = 13$

   c $2x - y = -4$

      $x - y = -5$

2 Solve each pair of simultaneous equations.

   a $x + 2y = 8$

      $2x + 5y = 17$

   b $3x + 3y = 6$

      $2x - 4y = 10$

   c $3x + 5y = 15$

      $x + 3y = 7$

3 A cafe sells a total of 500 sandwiches and baguettes in one day. They take a total of £2190 for the sandwiches and baguettes. Sandwiches cost £3.75 each and baguettes cost £4.75 each.

How many sandwiches were sold?

Hint: Write out your equations first – check that they make sense.

ALGEBRA

25

☑ Solving simultaneous equations: one linear, one non-linear

Use the method of substitution to solve the following pairs of equations.

1 State clearly the points of intersection of the straight-line graph and the quadratic graph in each case.

a $y = x^2 + 6x + 4$

$y = 2x + 1$

b $y = x^2 - 4x + 8$

$y = 16 - 2x$

_____

_____

_____

_____

_____

_____

_____

2 State clearly the points of intersection of the circle and the straight line in each case.

a $x^2 + y^2 = 29$

$y = x - 7$

b $x^2 + y^2 = 13$

$y - x = 1$

_____

_____

_____

_____

_____

_____

_____

3 A straight line has the equation $y = 3x + 2$

A curve has the equation $y^2 = -12(x + 1)$

Hint: Don't try to use a different method.

a Use the method of substitution to solve the simultaneous equations and find any points of intersection of the line and the curve.

_____

_____

_____

_____

_____

_____

_____

_____

_____

b Here are three sketches showing the curve $y^2 = -12(x + 1)$ and three possible positions of the line $y = 3x + 2$

Sketch 1

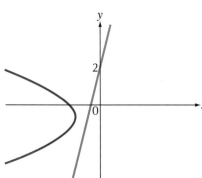

Sketch 2

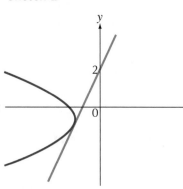

Sketch 3

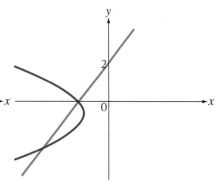

Which is the correct sketch? Give a reason for your choice.

_____

_____

_____

_____

✓ Factorising quadratic expressions

Factorise the following expressions.

1 $x^2 + 7x + 10$

$(x + 5)(x + 2)$

2 $x^2 + 10x + 21$

$(x + 7)(x + 3)$

3 $x^2 + 8x + 12$

$(x + 6)(x + 2)$

4 $x^2 + 10x + 25$

$(x + 5)(x + 5)$

5 $x^2 + x - 6$

$(x + 3)(x - 2)$

$x^2 - 2x + 3x - 6$

6 $x^2 + 2x - 15$

$(x + 5)(x - 3)$

7 $x^2 + 4x - 12$

$(x + 6)(x - 2)$

8 $x^2 + 6x - 16$

$(x + 8)(x - 2)$

9 $x^2 - 7x + 12$

$(x - 4)(x - 3)$

10 $x^2 - 8x + 15$

$(x - 5)(x - 3)$

11 $x^2 - x - 30$

$(x - 6)(x + 5)$

12 $x^2 - 7x - 18$

$(x - 9)(x + 2)$

ALGEBRA

Solving quadratic equations

Solve the following quadratic equations by factorising.

1 $x^2 + x - 6 = 0$

$(x + 3)(x - 2)$
$x = 2$ or -3

2 $x^2 - 8x + 15 = 0$

$(x \mp 5)(x - 3)$
$x = 5$ or 3

3 $x^2 + 5x + 6 = 0$

$(x + 3)(x + 2)$
$x = -3$ or -2

4 $x^2 + 8x + 15 = 0$

$(x + 5)(x + 3)$
$x = -5$ or -3

5 $x^2 + 5x - 14 = 0$

$(x + 7)(x - 2)$
$x = -7$ or 2

6 $x^2 - 6x - 72 = 0$

$(x - 12)(x + 6)$
$x = 12$ or -6

7 I think of a positive number. I square it, and add double the number. The answer is 15. What number did I think of? $2x + x^2 = 15$

3

8 I think of a negative number. I square it, and add triple the number. The answer is 10. What number did I think of?

$x^2 + 3x = 10$
$x^2 - 3x = 10$
-5

☑ Factorising harder quadratic expressions

Factorise the following quadratic expressions.

24.

1 $3x^2 + 5x + 2$

$(3x + 2)(x + 1)$

2 $6x^2 + 25x + 4$

$(6x + 6)(x + 4)$

$6x^2 + 24x^2 + 6x + 24$

$(6x + 4)(x + 1)$

$6x^2 + 6x + 4x + 4x$

$(6x + 1)(x + 4)$

3 $3x^2 + 10x - 8$

$3x^2 + 12x - 2x - 8$

$3x(x + 4) - 2(x + 4)$

$(3x - 2)(x + 4)$

4 $3x^2 + 16x - 12$

$3x^2 + 18x - 2x - 12$

$3x(x + 6) - 2(x + 6)$

$(3x - 2)(x + 6)$

5 $6x^2 - 13x + 6$

$6 \times 6 = 36$

$6 + 6 \quad 1 \quad 2$

$(6x + 3)(12x - 2)$

$6x^2 \quad 6x^2 - 12x - 8 + 6$

$6x(x - 2) + - (x + 6)$

$6x^2 - 9x - 4x + 6$

$3x(2x - 3) - 2(2x - 3)$

$(3x - 2)(2x - 3)$

6 $3x^2 - 11x + 10$

$3x^2 - 6x - 5x + 10$

$3x(x - 2) - 5(x - 2)$

$(3x - 5)(x - 2)$

7 $4x^2 - 4x - 3$

$4x^2 + 2x - 6x - 3$

$2x(2x + 1) - 3(2x + 1)$

$(2x - 3)(2x + 1)$

8 $3x^2 - 4x - 4$

$3x^2 - 6x + 2x - 4$

$3x(x - 2) + 2(x - 2)$

$(3x + 2)(x - 2)$

ALGEBRA

30

Solving harder quadratic equations: factorising

Solve the equations in questions **1–4** by factorising, giving answers as a fraction where necessary.

1 $2x^2 + 7x + 3 = 0$

$2x^2 + 6x + x + 3$

$2x(x+3) + 1(x+3)$

$(2x+1)(x+3)$

$x = -\frac{1}{2} \text{ or } -3$

2 $6x^2 + x - 2 = 0$

$6x^2 - 3x + 4x - 2$

$3x(2x-1) + 2(2x-1)$

$(3x+2)(2x-1).$

$x = -\frac{2}{3} \text{ or } \frac{1}{2}$

3 $10x^2 + x - 3 = 0$

$10x^2 - 5x + 6x - 3 = 0$

$5x(2x-1) + 3(2x-1)$

$(5x+3)(2x-1)$

$x = -\frac{3}{5} \text{ or } \frac{1}{2}$

4 $4x^2 - 29x + 7 = 0$

$4x^2 - x - 28x + 7 = 0$

$x(4x-1) - 7(4x-1).$

$(4x-1)(x-7)$

$x = \frac{1}{4} \text{ or } 7$

5 **a** Write down an algebraic expression for the area of the rectangle.

$(5x - 24)(3x) = 15x^2 - 72x$

$3x$

$5x - 24$

b The area of the rectangle is $15\,cm^2$. Form and solve an algebraic equation to find the value of x.

$15 = 15x^2 - 72x.$ $15x^2 - 72x - 15 = 0$

$15x^2 - 75x + 3x - 15 = 0$ $15x(x-5) + 3(x-5)$

$(15x+3)(x-5)$ $x = 5 \text{ or } -\frac{3}{15} / -\frac{1}{5}$

Solving quadratic equations:

using $x = \dfrac{-b \pm \sqrt{b^2 - 4ac}}{2a}$

Use the formula to solve the following quadratic equations, giving answers correct to 2 decimal places.

1 $x^2 + 4x + 1 = 0$

2 $x^2 - 3x - 2 = 0$

3 $x^2 + 10x + 1 = 0$

4 $5x^2 + 2x - 9 = 0$

5 $3x^2 - 2x - 9 = 0$

6 $4x + 3 - 2x^2 = 0$

Completing the square

1 Write each expression in the form $(x + p)^2 + q$, where p and q are integers or fractions.

a $x^2 + 12x + 7$

b $x^2 + 4x - 6$

c $x^2 - 8x + 6$

d $x^2 - 20x - 1$

e $x^2 + 7x + 6$

f $x^2 - 5x - 0.5$

2 Solve each quadratic equation by completing the square.

a $x^2 + 14x + 13 = 0$

b $x^2 + 2x - 8 = 0$

c $x^2 - 8x + 13 = 0$

d $x^2 + 3x - 8 = 0$

☑ Solving equations with algebraic fractions

Hint: As with ordinary fractions, make the denominators the same.

Solve the following equations with algebraic fractions.

1 $\dfrac{x}{5} - \dfrac{2x+4}{3} = 1$ $\quad \overline{15}$

$\dfrac{3x}{15} - \dfrac{10x+20}{15} = 1$

$\dfrac{-7x-20}{15} = 1$

2 $\dfrac{x+1}{3} - \dfrac{x-3}{2} = 1$

3 $\dfrac{x-2}{2} + \dfrac{x+1}{3} = 4$

4 $\dfrac{x-3}{5} - \dfrac{x-2}{3} = 5$

5 $\dfrac{x+2}{5} - \dfrac{2x-1}{4} = 4$

6 $\dfrac{2x-5}{3} - \dfrac{3x-8}{5} = 3$

Solving harder equations with algebraic fractions

Solve the following equations with algebraic fractions.

1 $\dfrac{2}{x+2} + \dfrac{3}{x+3} = 2$

2 $\dfrac{3}{x+3} + \dfrac{4}{x+4} = 5$

3 $\dfrac{9}{2x-7} - \dfrac{6}{x-1} = 2$

4 $\dfrac{12}{x-1} - \dfrac{8}{x+3} = 2$

5 $\dfrac{8}{2x-1} - \dfrac{2}{x+1} = 2$

6 $\dfrac{6}{x-1} + \dfrac{9}{2x-3} = 3$

ALGEBRA

Hint: Use the formula if you need to.

Solving linear inequalities

1 Solve each inequality.

a $\dfrac{x+3}{2} < 5$

_____

_____

_____

b $\dfrac{3x-6}{5} \geqslant 3$

_____

_____

_____

c $\dfrac{x}{10} + 2 \leqslant -5$

_____

_____

_____

d $8x + 4 > 6x - 2$

_____

_____

_____

e $6 - 2x \geqslant 2 - 4x$

_____

_____

_____

f $\dfrac{5-2x}{4} \leqslant 0.5$

_____

_____

_____

2 The area of this rectangle is greater than its perimeter.

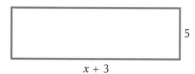

5

$x + 3$

a Write an inequality and solve it to find the range of values of x.

_____

_____

b Given that x is an integer, write down the smallest possible value of x.

_____

Simplifying algebraic fractions

1 Simplify the following fractions.

a $\dfrac{25a}{5}$ _____

b $\dfrac{20a^2}{4a}$ _____

c $\dfrac{16f^2}{32f^2}$ _____

d $\dfrac{12a^2b^3}{30a^3b^2}$ _____

e $\dfrac{8x + 16}{4}$ _____

f $\dfrac{2a - 6}{3a - 9}$ _____

g $\dfrac{12t + 20}{3t + 5}$ _____

h $\dfrac{2a + ab}{a^2}$ _____

i $\dfrac{f - 5}{(f - 5)^2}$ _____

j $\dfrac{(a + 2)^3}{(a + 2)^2}$ _____

2 Simplify the following expressions.

a $\dfrac{x^2 + 8x + 7}{x + 1}$ _____

_____

b $\dfrac{30x^2y^2 - 2x}{15xy^2 - 1}$ _____

_____

c $\dfrac{x^2 - 2x - 8}{x^2 + 5x + 6}$ _____

_____

d $\dfrac{x - 4}{x^2 - 16}$ _____

_____

e $\dfrac{2x^2 - 5x + 2}{6x^2 - 7x + 2}$ _____

_____

Simplifying algebraic fractions (addition and subtraction)

Hint: As with ordinary fractions, make the denominators the same.

1 Work out the following, simplifying where possible.

a $\dfrac{4a}{3} + \dfrac{3a}{5}$ _____

b $\dfrac{b-3}{4} - \dfrac{b+7}{5}$ _____

c $\dfrac{2c+4}{2} - \dfrac{3c-3}{5}$ _____

d $\dfrac{7}{d} + \dfrac{3}{e}$ _____

2 Work out the following, simplifying where possible.

a $\dfrac{5}{2a} + \dfrac{4}{3a^2}$ _____

b $\dfrac{3}{b-4} + \dfrac{3}{b+1}$ _____

c $\dfrac{1}{x-3} - \dfrac{2}{x-5}$ _____

d $\dfrac{10}{x-5} + \dfrac{2}{(x-5)^2}$ _____

3 Find and simplify an expression for the perimeter of the rectangle.

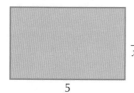

$\dfrac{3}{x-4}$

$\dfrac{5}{x+3}$

_____

_____

_____

_____

_____

ALGEBRA

Simplifying algebraic fractions (multiplication and division)

1 Work out the following, simplifying where possible.

a $\dfrac{x}{3} \times \dfrac{x}{5}$ _____

b $\dfrac{x}{3} \div \dfrac{y}{5}$ _____

c $\dfrac{x}{y^2} \times \dfrac{y^3}{5x}$ _____

d $\dfrac{x}{y^2} \div \dfrac{x}{5y}$ _____

e $\dfrac{80a}{16}$ _____

f $\dfrac{20a^2}{4a}$ _____

2 Work out the following, simplifying where possible.

a $\dfrac{x+2}{3} \times \dfrac{x-4}{5}$ _____

b $\dfrac{x^2-9x+18}{15} \div \dfrac{x^2-x-30}{25}$ _____

_____

c $\dfrac{t+5}{4} \times \dfrac{t^2-4t}{t^2+5t}$ _____

d $\dfrac{x^2}{x^2+3} \div \dfrac{x}{x+3}$ _____

3 Work out the following, simplifying where possible.

a $\dfrac{x^2-16}{7} \times \dfrac{21}{x^2-6x+8}$ _____

_____

b $\dfrac{x^2-4x}{3x^2+2x-8} \div \dfrac{3x^2-16x+16}{x^2-2x-8}$ _____

_____

_____

_____

☑ Proportionality

Hint: Remember the difference between the proportionality equations of direct and inverse proportionality.

1 x is directly proportional to y. When $x = 28$, $y = 7$.

   a Write the proportionality equation for x and y.

   _____

   _____

   b Work out x when $y = 3$. _____

   c Work out y when $x = 44$. _____

2 x is inversely proportional to y. When $x = 20$, $y = 5$.

   a Write the proportionality equation for x and y.

   _____

   _____

   b Work out x when $y = 10$. _____

   c Work out y when $x = 25$. _____

3 The area of a circle is directly proportional to the square of its radius.

When the radius of a circle is 5 cm, its area is 78.5 cm^2.

By first writing out the proportionality equation for area and radius, work out the area of a circle with a radius of 7 cm.

_____

_____

4 The cooking time in a microwave oven is assumed to be inversely proportional to its power. The five power settings on a microwave oven are shown in the table on the right.

   a A jacket potato takes 8 minutes on 'Full'. How long would it take on 'Heat'?

   _____

   b A large bowl of soup takes 12 minutes on 'Simmer'. How long will it take on 'Heat'?

   _____

Level	Power (W)
Full	800
Heat	600
Simmer	350
Defrost	200
Warm	150

ALGEBRA

5 The force of attraction, *F*, between two magnets is inversely proportional to the square of the distance, *d*, between them. When the magnets are 5 cm apart, the force of attraction is 22 newtons. How far apart are they when the force of attraction is 88 newtons?

_____

_____

_____

_____

6 Match each statement to a table.

Table A

x	1	2	3
y	0.5	2	4.5

a *y* is inversely proportional to *x*.

Table _____

Table B

x	1	2	3
y	5	2.5	1.66 (3 sf)

b *y* is directly proportional to x^2.

Table _____

Table C

x	1	2	3
y	10	2.5	1.11 (2 dp)

c *y* is inversely proportional to x^2.

Table _____

7 Match each graph with the corresponding proportion.

a $y \propto x$

b $y \propto x^2$

c $y \propto \dfrac{1}{x}$

Sketch 1

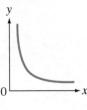

Sketch 2

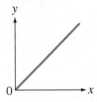

Sketch 3

_____   _____   _____

✅ Arcs and sectors

1 Calculate the arc length of each of these sectors, to 3 significant figures.

a

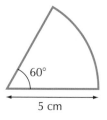

60°

5 cm

b

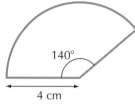

140°

4 cm

c

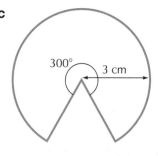

300° 3 cm

_____ _____ _____

_____ _____ _____

_____ _____ _____

2 Calculate the area of each of these sectors, giving answers in terms of π.

a

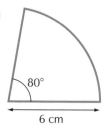

80°

6 cm

b

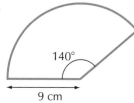

140°

9 cm

c

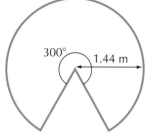

300° 1.44 m

_____ _____ _____

_____ _____ _____

_____ _____ _____

3 Carly plans to lay a patio in the corner of her garden.

a What is the area of her patio? Give the answer to 2 decimal places.

_____

_____

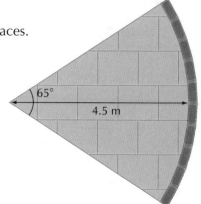

65°

4.5 m

She wants to add an edging of single bricks to the patio.
Each brick is 20 cm long.

b How many bricks does she need to lay?

_____

_____

Surface area of cylinders, cones and spheres

Hint: Surface area of cylinder = $2\pi rh + 2\pi r^2$, of cone = $\pi r^2 + \pi rl$, of sphere = $4\pi r^2$

Calculate the surface area of each of the following solid shapes, correct to 1 decimal place.

1

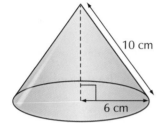

10 cm

6 cm

_____

_____

_____

2

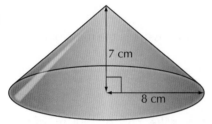

7 cm

8 cm

_____

_____

_____

_____

_____

_____

3

2.8 m

3.2 m

_____

_____

_____

_____

4

11 mm

_____

_____

_____

5

26 cm

_____

_____

_____

GEOMETRY AND MEASURES

 **Density**

1 Oak has a density of $0.75\,\text{g/cm}^3$. Find the mass of $1000\,\text{cm}^3$ of oak.

_____

2 $500\,\text{cm}^3$ of pine has a mass of 19 grams. Find the density of the pine.

_____

3 A 1 kg bag of sugar has a volume of approximately $620\,\text{cm}^3$. What is the density of sugar?

_____

4 An average hot-air balloon contains $2900\,\text{m}^3$ of air.

a Hot air in a hot-air balloon has a density of around $0.9\,\text{kg/m}^3$. What is the mass of the hot air in a hot-air balloon?

_____

b Air at normal temperature has a density of $1.2\,\text{kg/m}^3$. What is the mass of $2900\,\text{m}^3$ of this air?

_____

c The 'lift' of a hot-air balloon is the difference in mass between the hot air inside the balloon and the equivalent volume of air at normal temperature. What is the lift of an average hot-air balloon?

_____

5 The iceberg that sank the *Titanic* had a volume of about $500\,000\,\text{m}^3$. Ice has a density of $0.92\,\text{g/cm}^3$. What was the mass of the iceberg, to the nearest tonne?

_____

_____

6 The density of gold is $19.3\,\text{g/cm}^3$. Find the side length of a cube of gold that has a mass of $521.1\,\text{kg}$.

_____

_____

7 The density of hydrogen is $0.0899\,\text{kg/m}^3$. Find the side length of a cube of hydrogen that has a mass of $521.1\,\text{kg}$.

_____

_____

✓ Volume of cones and spheres

1 Calculate the volume of each of these solid shapes, giving the answer in terms of π.

> **Hint:** Volume of cone = $\frac{1}{3}\pi r^2 h$, of sphere = $\frac{4}{3}\pi r^3$

a

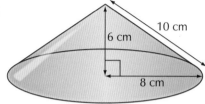

_____

_____

_____

b

_____

_____

_____

2 A football of diameter 27 cm is packed into a cubical box of side length 28 cm.

How much empty space is in the box? Give your answer to 3 significant figures.

_____

_____

_____

3 Three spheres of radius 3 cm just fit into a closed cylinder.

How much empty space is there in the cylinder? Give your answer to the nearest cm³.

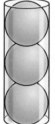

_____

_____

_____

_____

4 A wooden cone of perpendicular height 15 cm and base radius 12 cm has its top 5 cm cut off to leave a frustum. The wood has a density of 0.78 g/cm³. Find the mass of the frustum, to the nearest gram.

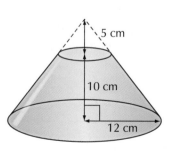

_____

_____

_____

Volume and surface area

1 The following shape is made from a cylinder and a cone.

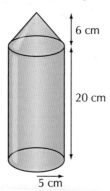

6 cm

20 cm

5 cm

Find the volume of the shape, in terms of π.

_____

_____

_____

_____

_____

2 The following shape is made from a cylinder and a hemisphere.

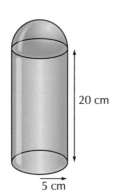

20 cm

5 cm

Find the surface area of the shape, in terms of π.

_____

_____

_____

_____

_____

3 **a** An artist makes a concrete statue in the shape of a cylinder with a cone on top. The concrete has a density of 2.4 g/cm^3.

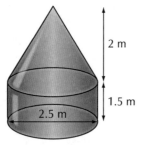

2 m

1.5 m

2.5 m

What is the mass of the statue, to the nearest kilogram?

_____

_____

_____

b The artist plans to cover the surface of his statue with gold leaf. Gold leaf costs £18.95 for a book of 25 sheets, each measuring 80 mm by 80 mm. How much will it cost to cover the statue with gold leaf?

_____

_____

_____

Similar shapes

1 The diagram shows a picture frame containing a mount.

Are the two rectangles similar? Explain your answer.

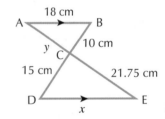

25 cm

20 cm

19 cm 14 cm

_____

_____

2 The diagrams below show two similar triangles.

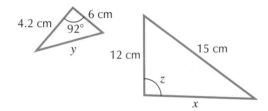

4.2 cm 6 cm

92°

y

12 cm 15 cm

z

x

Find the value of x, y and z.

a $x =$ _____

b $y =$ _____

c $z =$ _____

3 In the diagram on the right, AB is parallel to DE.

AB = 18 cm, BC = 10 cm, CD = 15 cm and CE = 21.75 cm

a Explain why triangles ABC and EDC are similar.

18 cm

A B

y C

10 cm

15 cm 21.75 cm

D E

x

_____

_____

b Calculate the lengths x and y.

i $x =$ _____ **ii** $y =$ _____

4 A mobile phone mast is 18 m high. At 11 am it casts a shadow 32 m long. At the same time, an electricity pylon near to the mast casts a shadow 56 m long. Find the height of the pylon.

Hint: A quick sketch will probably help.

_____

_____

_____

✅ Scale factors

1 A cuboid is 4 cm by 8 cm by 10 cm. A similar cuboid is 12 cm by 24 cm by 30 cm.

    a What is the linear scale factor? _____

    b What is the area scale factor? _____

    c What is the volume scale factor? _____

2 A photograph is being enlarged by a linear scale factor of 2. The area of the original photo is 32 square inches. What is the area of the enlargement?

_____

3 A photograph is being enlarged by a linear scale factor of $\frac{2}{3}$. The area of the original photo is 2700 cm^2. What is the area of the enlargement?

_____

4 These two rectangles are similar, with areas of 129 cm^2 and 21 cm^2. Find the length marked x. Give your answer to 1 decimal place.

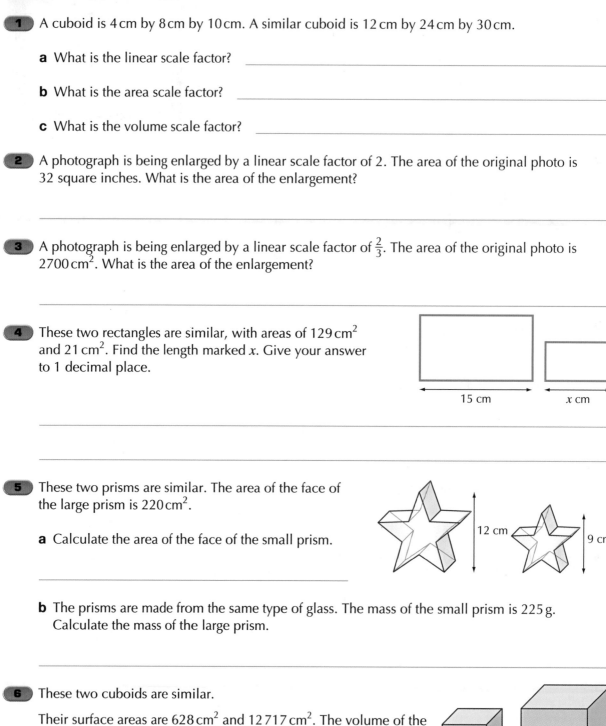

15 cm x cm

_____

_____

5 These two prisms are similar. The area of the face of the large prism is 220 cm^2.

12 cm   9 cm

    a Calculate the area of the face of the small prism.

_____

    b The prisms are made from the same type of glass. The mass of the small prism is 225 g. Calculate the mass of the large prism.

_____

6 These two cuboids are similar.

Their surface areas are 628 cm^2 and 12 717 cm^2. The volume of the smaller cuboid is 1040 cm^3. Find the volume of the larger cuboid.

_____

_____

_____

2D trigonometry

1 A ladder 5 m long rests against a wall.

The foot of the ladder is 2.2 m from the base of the wall.

What angle does the ladder make with the wall? Give your answer to 3 significant figures.

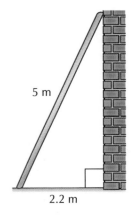

5 m

2.2 m

_____

_____

2 Freda is 4.5 km from the world's largest wind turbine. She measures the angle of elevation to its tip as 2.5°. How tall is the wind turbine, to the nearest metre?

Hint: A quick sketch will probably help.

_____

_____

3 Find the size of angle x, correct to 2 decimal places.

Hint: Find the vertical height first.

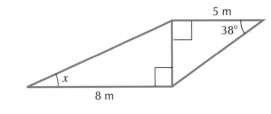

5 m

38°

x

8 m

_____

_____

_____

_____

4 Find the size of the acute angle made between the two diagonals of a rectangle of sides 10 cm and 6 cm. Give your answer to 3 significant figures.

_____

_____

_____

_____

5 A ship sails 40 km on a bearing of 055°. How far north has it travelled?

> **Hint:** A quick sketch will probably help.

_____

_____

6 By first finding its perpendicular height, calculate the area of the isosceles triangle.

_____

_____

_____

_____

_____

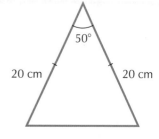

20 cm   50°   20 cm

7 An aeroplane is flying at a constant height of 10 000 m. An observer in the plane measures the angle of depression to a point on the ground as 30°. Twenty seconds later the angle of depression to the same point on the ground is 35°.

a Work out the horizontal distance travelled by the plane in 20 seconds.

_____

_____

_____

_____

b What is the speed of the plane, in km/h?

_____

_____

3D Pythagoras' theorem and trigonometry problems

Give answers to 3 significant figures where appropriate.

1 The diagram on the right shows a cube with a side length 10 cm.

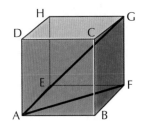

a Calculate the length of AF.

_____

_____

b Calculate the length of AG.

_____

_____

c Calculate the size of angle FAG.

_____

_____

2 The diagram on the right shows a cuboid.

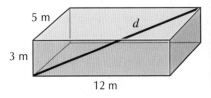

a Calculate the length of the diagonal, *d*.

_____

_____

_____

_____

b Calculate the angle this diagonal makes with the base of the cuboid.

_____

_____

_____

3 The diagram on the right shows a wedge.

AB = 10 cm, BD = 3 cm and DF = 9 cm

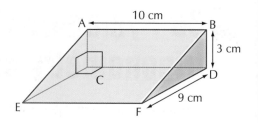

a Find the length of EB.

_____

_____

_____

_____

b Find the size of angle BFD.

_____

_____

c Find the size of angle AFC.

_____

_____

4 The diagram on the right shows a rectangular-based pyramid.

V is vertically above the centre of the rectangle.

AB = CD = 10 cm, AD = BC = 12 cm
AV = BV = CV = DV = 20 cm

Calculate the angle AV makes with the base ABCD.

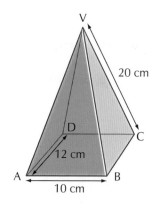

_____

_____

_____

_____

_____

_____

_____

Sine rule and cosine rule

Sine rule: $\dfrac{a}{\sin A} = \dfrac{b}{\sin B} = \dfrac{c}{\sin C}$ Cosine rule: $a^2 = b^2 + c^2 - 2bc \cos A$

1 Calculate the value of x in each of these triangles.

a

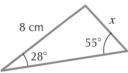

8 cm
x
55°
28°

b

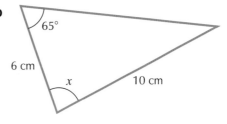

65°
6 cm
x
10 cm

_____   _____

_____   _____

2 Two ships leave port at the same time. Ship A travels at a constant speed of 14 km/h on a bearing of 065°. Ship B travels at a constant speed of 21 km/h on a bearing of 135°. Calculate the distance between ships A and B after one hour.

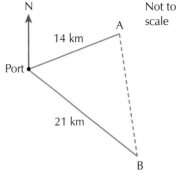

N   Not to scale
14 km   A
Port
21 km
B

_____

_____

_____

_____

_____

3 Rajesh stands near an office block and measures the angle of elevation to the top of the building as 68°. He then moves 50 m further away and measures the angle of elevation as 43°.

Hint: Find all the angles first.

How tall is the office block?

Not to scale
43° 68°
50 m

_____

_____

_____

_____

_____

_____

4 The lengths of the adjacent sides of a parallelogram are 10 cm and 15 cm. The longer diagonal is 19 cm. Work out the length of the shorter diagonal.

> **Hint:** A quick sketch will probably help.

5 A boat sails from port 20 km due east to a lighthouse. It then changes course and sails 50 km on a bearing of 150° to a buoy. The boat then sails directly back to port.

a How far does the boat sail altogether?

b On what bearing does the boat sail back from the buoy to the port?

Areas of triangles

A farmer has several triangular fields.

1 Find the area of this field, to 3 significant figures.

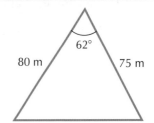

_____

_____

2 The farmer estimates the area of this field as $34\,500\,m^2$.

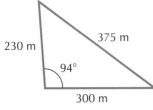

Find the difference between the actual area of the triangular field and her estimate.

_____

_____

_____

_____

3 Calculate the area of the field ABCD.

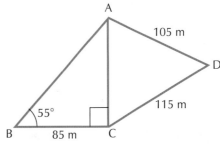

_____

_____

_____

_____

_____

_____

_____

_____

_____

_____

Congruent triangles

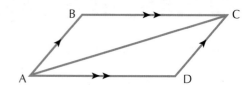

1 ABCD is a parallelogram.

Prove that triangle ABC is congruent to triangle CDA.

BC = AD

AB = DC

AC is common

so congruent

2 In the diagram, AB is equal in length and parallel to DE.

Prove that triangle ABC is congruent to triangle EDC.

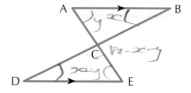

AB = DE

Same angles so same length as well

3 ABCD is a quadrilateral with AB = AD and BC = CD.

Prove that angle ABC = angle ADC.

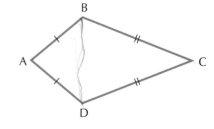

4 ABCDE is a regular pentagon.

Prove that triangle ABC is congruent to triangle CDE.

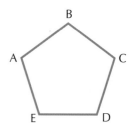

Circle theorems

1 Work out the sizes of the lettered angles. Give reasons in your answers.

a

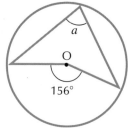

a

O

156°

$\underline{156/2 = 78}$

b

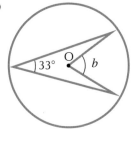

33° O b

66

c

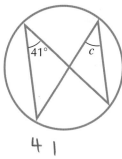

41° c

41

d

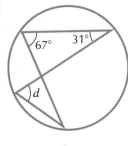

67° 31°

d

67

e

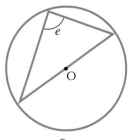

e

O

90

f

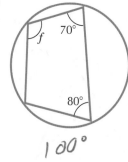

f 70°

80°

100°

g

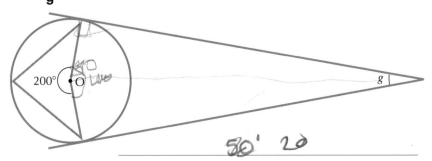

200° O

g

50° 20

h

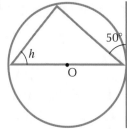

h

50°

O

i

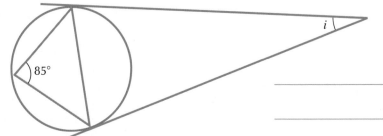

85°

i

Circle theorems

j

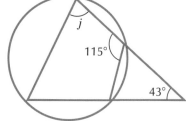

k

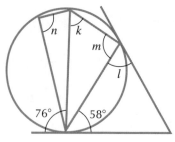

_____
_____
_____
_____
_____

_____
_____
_____
_____
_____

2 Prove that the angle at the centre is twice the angle at the circumference, **drawn from the same arc**.

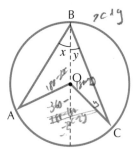

_____
_____
_____
_____

3 Prove that the angles **drawn from the same arc in the same segment** are equal.

_____
_____
_____
_____
_____

4 Prove that the **angle in a semicircle** is a right angle.

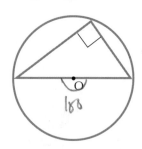

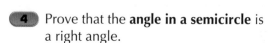

Angle at centre
is dable of
angle at circumference.

_____
_____

Enlargement

1 On the grid below, draw the enlargement of each object about its centre, O.

Use the given scale factor for each shape.

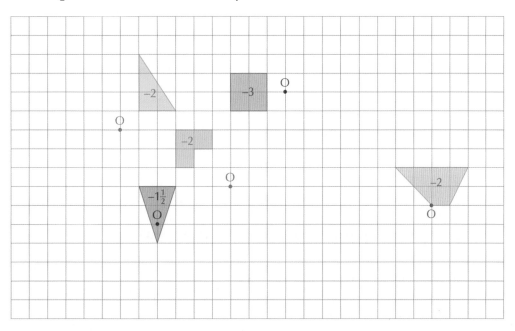

2 On the grid below, draw the enlargement of triangle A:

a by scale factor –1, centre (2, 0) and label the image B

b by scale factor –2, centre (0, 0) and label the image C

c by scale factor –$\frac{1}{2}$, centre (–5, 3) and label the image D

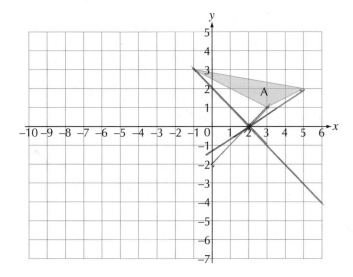

Enlargement

3 Enlarge the given rectangle by scale factor −2 with a centre of:

a (9, 6)   **b** (12, 5)

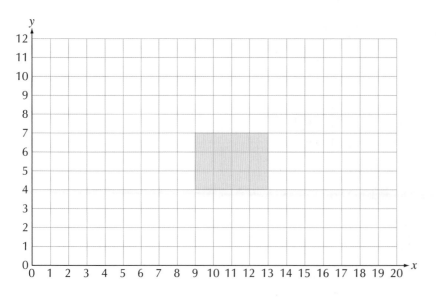

4 Triangle ABC has been enlarged to give triangle A'B'C'.

a What is the scale factor of the enlargement?

_____

b Draw construction lines to show the position of the centre of enlargement.

c Give the coordinates of the centre of enlargement.

_____

_____

d What is the inverse transformation that would take A'B'C' back to ABC?

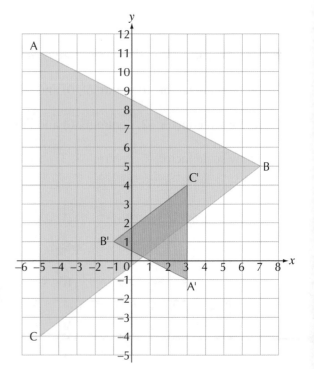

_____

_____

Vectors

1 $q = \begin{pmatrix} 1 \\ 3 \end{pmatrix}$ $r = \begin{pmatrix} -2 \\ -1 \end{pmatrix}$ $s = \begin{pmatrix} 4 \\ -2 \end{pmatrix}$

a On the square grid below, draw and label the vectors: **q**, **r**, **s**, $\frac{1}{2}$**s**, 2**q**, 3**r** and –**s**.

b Write the following as column vectors.

i –3**q**   **ii** 3**r**   **iii** $\frac{3}{2}$**s**

_____   _____   _____

2 $q = \begin{pmatrix} 1 \\ 3 \end{pmatrix}$ $r = \begin{pmatrix} -2 \\ -1 \end{pmatrix}$ $s = \begin{pmatrix} 4 \\ -2 \end{pmatrix}$

a On the square grid below, draw diagrams to illustrate these vectors.

q + r   **r – s**   **s + 2r**   **q + s**   **s – r**   **2q + 4r**

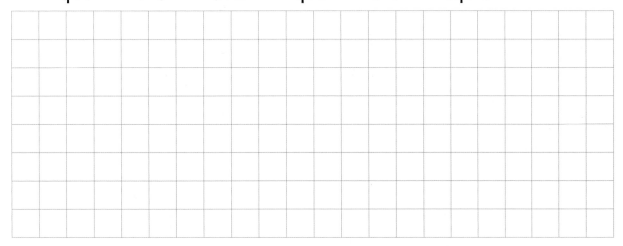

b Write the following as column vectors.

i **r + s**   **ii** **q – s**   **iii** **2q + s**

_____   _____   _____

3 In the parallelogram grid, $\overrightarrow{OA}$ = **a** and $\overrightarrow{OB}$ = **b**.

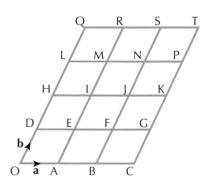

a Name two vectors equivalent to **a** _____ _____

b Name two vectors equivalent to **b** _____ _____

c Name two vectors equivalent to –**a** _____ _____

d Name two vectors equivalent to –**b** _____ _____

Give the following vectors in terms of **a** and **b**.

e $\overrightarrow{OC}$ = _____ **f** $\overrightarrow{FH}$ = _____ **g** $\overrightarrow{OT}$ = _____

h $\overrightarrow{AN}$ = _____ **i** $\overrightarrow{IK}$ = _____ **j** $\overrightarrow{NC}$ = _____

4 Identify and label the points C to J on the grid below.

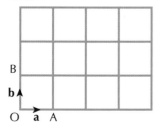

a $\overrightarrow{OC}$ = 3**a**   **b** $\overrightarrow{OD}$ = 2**b**   **c** $\overrightarrow{OE}$ = 3**a** + **b**

d $\overrightarrow{OF}$ = 2**a** + 3**b**   **e** $\overrightarrow{OG}$ = 4**a** + $\frac{3}{2}$**b**   **f** $\overrightarrow{OH}$ = $\frac{5}{2}$(**a** + **b**)

g $\overrightarrow{IO}$ = –3**a** – 2**b**   **h** $\overrightarrow{JO}$ = –$\frac{3}{2}$**a** – $\frac{5}{2}$**b**

5 $\overrightarrow{OA}$ = **a**, $\overrightarrow{OB}$ = **b** and M is the midpoint of AB.

Give the following vectors in terms of **a** and **b**.

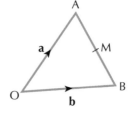

a $\overrightarrow{AB}$ = _____ **b** $\overrightarrow{AM}$ = _____ **c** $\overrightarrow{BA}$ = _____

e $\overrightarrow{BM}$ = _____ **f** $\overrightarrow{OM}$ = _____ **g** $\overrightarrow{MO}$ = _____

6 The diagram shows a regular hexagon ABCDEF with centre O.

Express the following vectors in terms of **a** and/or **b**.

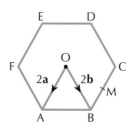

a $\overrightarrow{AB}$ = _____   **b** $\overrightarrow{DC}$ = _____   **c** $\overrightarrow{FE}$ = _____

M is the midpoint of BC.

d $\overrightarrow{EM}$ = _____

Cumulative frequency diagrams

1 The lengths of time that 100 people spent waiting to see their doctor are shown in the table.

Time, t (mins)	$0 < t \leqslant 5$	$5 < t \leqslant 10$	$10 < t \leqslant 15$	$15 < t \leqslant 20$	$20 < t \leqslant 25$	$25 < t \leqslant 30$
Frequency	18	7	35	22	11	7

a Draw a cumulative frequency diagram to illustrate this data.

Cumulative frequency diagrams

> **Hint:** Draw lines on your diagram to help you answer the questions. Use a ruler to make sure you are accurate.

b Use your cumulative frequency diagram on page 63 to estimate:

  i the median time _____

  ii the lower quartile _____

  iii the upper quartile _____

  iv the interquartile range _____

c How many people waited up to 12 minutes? _____

d How many people waited less than 22 minutes? _____

e How many people waited between 14 and 24 minutes? _____

2 The cumulative frequency diagram shows the mass of a random sample of 80 eggs from a day's production on a farm.

  a Use the cumulative frequency diagram to estimate the interquartile range.

_____

_____

  b Use the cumulative frequency diagram to find the number of eggs that weigh between 55 and 60 grams.

_____

_____

  c The farmer only sells the eggs that weigh at least 45 grams. What percentage of the eggs does he not sell?

_____

_____

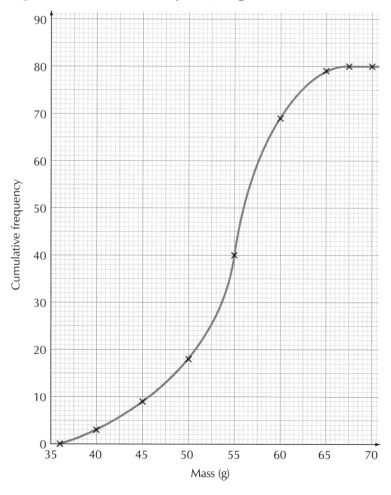

Box plots

1 In each of the box plots below, find:

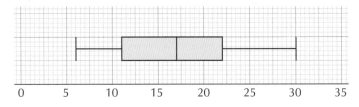

a **i** the median _____

    ii the range _____

    iii the interquartile
        range _____

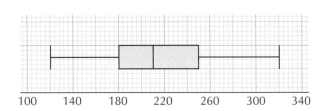

b **i** the median _____

    ii the range _____

    iii the interquartile
        range _____

2 The box plot shows the time taken by a group of 15-year-old boys to solve a puzzle.

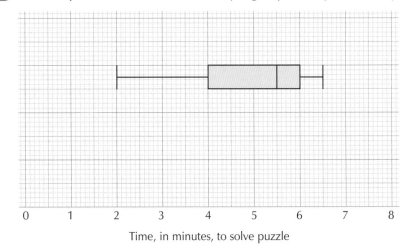

Time, in minutes, to solve puzzle

> **Hint:** Comparing the smallest and then the biggest will only count as one comparison, as the biggest and smallest are similar in the two plots.

The same puzzle was given to a group of 15-year-old girls. Their results are: shortest time 3 minutes 30 seconds, median 4 minutes 30 seconds, upper quartile 5 minutes 30 seconds, the range 4 minutes and the interquartile range 1 minute 30 seconds.

a Draw the girls' results as a box plot on the grid.

b Write four comparisons between the two distributions.

_____

_____

_____

_____

✓ Tree diagrams

1 **a** A 10p coin and a 50p coin are flipped together. Draw a tree diagram to show all the possible outcomes.

Hint: Remember to put the probabilities along each 'branch'.

        10p   **50p**

b Use the tree diagram to find the probability of each of the following events.

  i two heads _____

  ii one head then one tail _____

  iii one head and one tail _____

  iv at least one head _____

2 Bag A contains five toffees and five chocolates. Bag B contains five toffees and two chocolates. Mo takes a sweet at random from each bag.

  a Draw a tree diagram to show all the possible outcomes.

        A   **B**

Use your tree diagram to find the probability that:

b she picks two toffees _____

c she picks at least one toffee _____

d she does not pick a toffee _____

3 Hugh, Jordan and Gemma are taking a maths examination. Their maths teacher estimates their chances of getting an A grade to be 0.95, 0.8 and 0.4 respectively.

a Draw a tree diagram to show all the possible outcomes.

Use the tree diagram to find the probability that:

b all three get an A grade _____

c Hugh and Jordan get an A grade, but Gemma does not _____

d at least one gets an A grade _____

e any two get an A grade _____

4 Amy has a password on her work computer that she is not very good at remembering. The probability that she gets it right first time is 0.85.

If she gets it wrong first time, the probability that she gets it right second time is 0.5.

If she gets it wrong the second time, the probability that she gets it right third time is 0.3.

If she gets it wrong three times in a row she has to phone the IT helpdesk to ask for the password.

What is the probability that she does not need to phone the IT helpdesk? _____

 Histograms

1 The table shows the heights of 100 Year 7 girls.

Height, h (cm)	$100 \leqslant h < 110$	$110 \leqslant h < 125$	$125 \leqslant h < 135$	$135 \leqslant h \leqslant 150$	$150 \leqslant h < 155$
Class width					
Frequency	8	33	37	18	4
Frequency density					

a Draw the histogram for this grouped data.

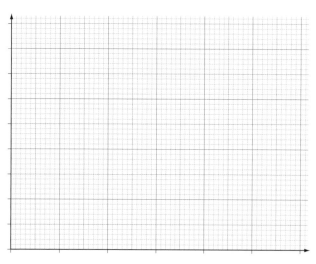

> **Hint:** Always put the frequency density along the vertical axis (the y-axis).

b Estimate the number of Year 7 girls who are less than 1.4 m tall.

_____

c Estimate the median height.

_____

d Find the interquartile range.

_____

e Estimate the mean height of the girls.

_____

_____

_____

2 The table shows the heights of 100 Year 7 boys.

Height, h (m)	$0.9 \leqslant h < 1.05$	$1.05 \leqslant h < 1.2$	$1.2 \leqslant h < 1.3$	$1.3 \leqslant h < 1.35$	$1.35 \leqslant h < 1.5$
Frequency	21	30	30	10	9

a Draw the histogram for this grouped data.

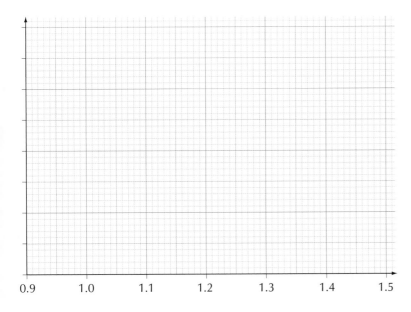

0.9   1.0   1.1   1.2   1.3   1.4   1.5

b Estimate the number of Year 7 boys who are taller than 1 m 10 cm.

_____

c Estimate the median height.

_____

d Find the interquartile range.

_____

_____

e Estimate the mean height of the boys.

_____

_____

3 The incomplete table and histogram below give some information about the mass of 80 cherry tomatoes.

Mass, **m** (grams)	$3 \leq m < 4$	$4 \leq m < 6$	$6 \leq m < 8$	$8 \leq m < 11$	$11 \leq m < 15$
Frequency	2	6			12

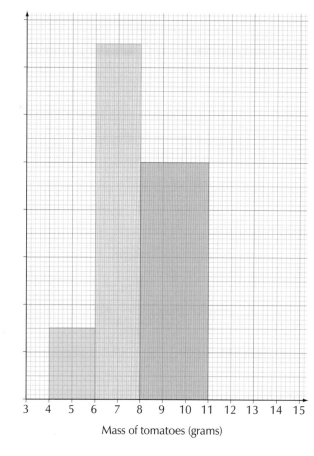

Mass of tomatoes (grams)

a Cherry tomatoes are classified as too large if they weigh more than 12 grams. Estimate the number of tomatoes that are too large.

_____

_____

b Cherry tomatoes are classified as too small if they weigh less than 4.5 grams. Estimate the number of tomatoes that are too small.

_____

_____

Stratified samples

This table shows the numbers of students in each year group of a school.

Year group	Girls	Boys	Total
7	124	126	250
8	127	132	259
9	136	151	287
10	144	156	300
11	144	170	314
Total	**675**	**735**	**1410**

1 The head teacher wants to survey 100 of these students.

a Which of the three methods below would give a random sample?

i Asking four Year 8 maths classes with 25 students in each class

ii Asking the first 100 students who arrive at school in the morning

iii Putting students in alphabetical order by name, then giving each a number from 1 to 1410
Putting tickets numbered 1 to 1410 into a bucket, then picking out 100 tickets

b Explain your choice and give a reason why the other methods would not be suitable.

> **Hint:** Don't just write, for example, 'It will be biased.' You must explain *why* it will be biased.

2 The head of mathematics decides that a stratified sample of 10% would be better. How many students from each year group should he survey?

☑ Probability: combined events

1 Bim shuffles a pack of cards and picks one at random. He replaces the card, shuffles the pack, then picks another card.

  a What is the probability that the first card Bim takes is a four?

  _____

  b What is the probability that the second card Bim takes is not a four?

  _____

  c Work out the probability that:

  i both cards are fours _____

  ii neither card is a four _____

  iii at least one card is a four _____

2 A bag contains 10 coloured balls. Six are red, three are green and one is gold. Jess takes a ball at random, notes its colour and replaces it in the bag. She then takes out another ball and notes its colour. Write down the probability that:

  a the first ball is red and the second ball is green _____

  b the first ball is green and the second ball is red _____

  c both balls are gold _____

  d neither ball is green _____

  e at least one ball is red _____

3 Richard, Steve and Tom take an Army entrance examination.

  The probabilities that they pass the written examination, the mental examination and the physical examination are shown in the table.

	Written exam	Mental exam	Physical exam
Richard	0.9	0.4	0.8
Steve	0.7	0.7	0.7
Tom	0.8	0.9	0.5

Work out the probability that:

a all three pass the physical exam _____

b all three fail the written exam _____

c Richard passes all three exams _____

d two pass but one fails the mental exam _____

4 A box of chocolates contains 10 caramel-centred chocolates and six toffee-centred chocolates. Craig takes two chocolates at random and eats them.

Work out the probability that the first chocolate was caramel-centred and the second was toffee-centred.

> **Hint:** How many chocolates are left after one has been eaten?

_____

_____

_____

5 Harij has four black socks and six blue socks in a drawer. He takes two socks out in the dark. What is the probability that he has a matching pair of socks?

_____

_____

_____

6 Bag A contains five 10p coins and three 2p coins. Bag B contains two 10p coins and six 2p coins.

Jim takes a coin at random from bag A and puts it into bag B. Then he takes a coin at random from bag B and puts it into bag A.

Work out the probability that bag A now has more 10p coins than 2p coins.

_____

_____

_____

_____

_____

STATISTICS AND PROBABILITY

Assessing understanding and problem solving

1 A train of length 175 m is travelling at 114 km/h as it approaches a tunnel of length 1.25 km.

How long will it take the train to pass completely through the tunnel at this velocity?

_____

_____

_____

(3 marks)

2 The diagram shows two right-angled triangles ABC and CDE.

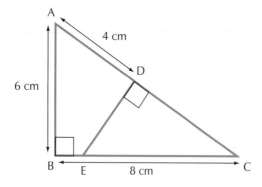

Find the length of the line DE.

_____

_____

_____

_____

(5 marks)

3 The diagram shows a square, of side length x. Inside the square is a shaded triangle with a vertex at a perpendicular distance y from the top edge.

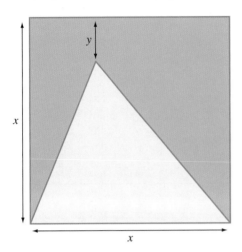

If $y = \frac{1}{4}x$, calculate the fraction of the square that is shaded.

_____

_____

_____

_____

_____

_____

_____

_____

_____

_____

_____

_____

_____

_____

_____

_____

(4 marks)

4 The diagrams show a workbench with two identical wooden blocks.

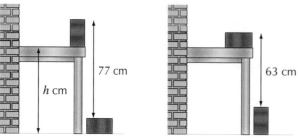

77 cm

h cm

63 cm

Calculate the height, *h*, of the workbench.

_____

_____

_____

_____

_____

_____

_____

_____

_____

_____

_____

_____

_____

_____

_____

_____

_____

_____

(7 marks)

5 A glass test tube is made from a cylinder and a hemisphere, as shown.

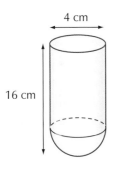

4 cm

16 cm

a Work out the total volume of the test tube. Give your answer in terms of π.

_____

_____

_____

_____

_____

(5 marks)

The test tube is filled with a chemical to a depth d cm, as shown below.

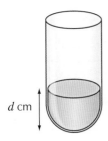

d cm

The chemical occupies exactly one-third of the full capacity of the test tube.

b Work out the value of d, to the nearest millimetre.

_____

_____

_____

_____

_____

(6 marks)

Assessing understanding and problem solving

6 Alex is working with this equation: $y = \dfrac{10x - 5}{\sqrt{16 - 9x^2}}$

Alex wants to draw a graph of the equation.

a Alex thinks the graph of the equation will only intersect the x-axis once.
Explain why Alex is correct.

_____

_____

_____

_____

_____

_____

_____

_____

(2 marks)

b Explain why Alex cannot find a value for y when x is $\frac{4}{3}$.

_____

_____

_____

_____

_____

_____

_____

(2 marks)

c Write down another value of x for which a value of y cannot be found.

_____

(1 mark)

7 In the diagram, BD = 9CD and AB = 5CD.

Show that $\sin A = \dfrac{k}{\sqrt{125}}$ and give the value of k.

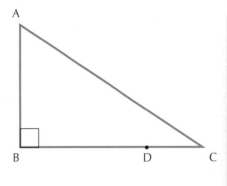

_____

_____

_____

_____

_____

_____

_____

_____

_____

(5 marks)

8 ABCDEFGH is a regular octagon, of side length 1 cm.

Calculate the exact length of AD.

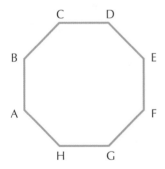

_____

_____

_____

_____

_____

_____

(7 marks)

Assessing understanding and problem solving

9 'Green garden' paint is made by mixing blue and yellow paint in the ratio 3 : 1.

'Green gold' paint is made by mixing blue and yellow paint in the ratio 1 : 3.

One litre of Green garden paint is mixed with half a litre of Green gold by mistake.

How much blue paint needs to be added to the mixture to make it Green garden again?

_____

_____

_____

_____

(6 marks)

10 Here is a trapezium-shaped tile.

30 cm

24 cm

Five of these tiles are arranged inside a blue rectangle that measures 24 cm by 30 cm.

Calculate the area of the blue rectangle that is still visible.

_____

_____

_____

_____

_____

_____

_____

_____

_____

_____

_____

_____

_____

(11 marks)

11 The sketch shows three identical rectangles with their sides parallel to the axes.

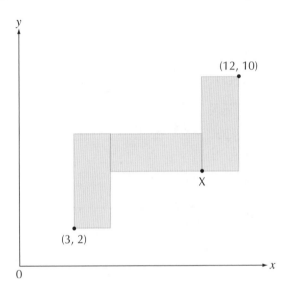

Find the coordinates of point X.

_____

_____

_____

_____

_____

_____

_____

(6 marks)

Assessing understanding and problem solving

12 Find the coordinates of the point where these two lines would meet if they were extended.

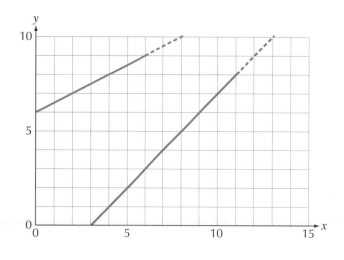

_____

_____

_____

_____

_____

_____

(8 marks)

13 A hot chocolate machine dispenses 155 millilitres of hot chocolate into cups with a capacity of 190 millilitres.

These values are accurate to 3 significant figures.

Milk is supplied in small cartons that contain 16 millilitres, accurate to the nearest millilitre.

Jenna likes her drinks to be milky and always adds two small cartons of milk to her hot chocolate.

Will Jenna's cup overflow?

You **must** show your working.

_____

_____

_____

_____

(4 marks)

14 Mali is using the quadratic formula to solve a quadratic equation.

After correctly substituting the values, he writes:

$$x = \frac{8 \pm \sqrt{64 - 96}}{6}$$

a What is the quadratic equation Mali is trying to solve?

_____

_____

(5 marks)

b Explain why Mali will not be able to find any solutions to the equation.

_____

_____

(1 mark)

15 Abbie has drawn a sketch of one of her kitchen wall and of one of the tiles she will use to tile it.

The wall has been measured to the nearest 10 cm and the tiles are measured to the nearest 0.5 cm.

Abbie will throw away any partly-used tiles, without trying to patch them.

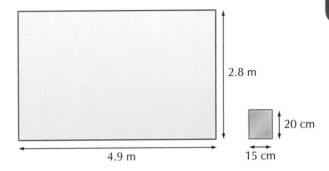

2.8 m

20 cm

4.9 m

15 cm

a Work out the minimum number of tiles Abbie will need.

_____

_____

_____

(5 marks)

b How many more tiles than the minimum might she need?

_____

_____

_____

(5 marks)

16 Five cones A, B, C, D and E are mathematically similar.

Their heights are in the ratio 2 : 3 : 4 : 5 : 6

Cone B has height 3.6 cm and surface area 45 cm^2.

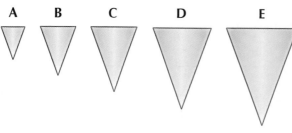

a Work out the height of cone C.

_____

_____

_____

_____

(2 marks)

b Work out the surface area of cone D.

_____

_____

_____

_____

(3 marks)

c Cone A is used to fill cone E with sand.

How many full cones of sand from cone A are needed to fill cone E?

_____

_____

_____

_____

(3 marks)

17 A slice of a circular 10-inch deep-pan pepperoni pizza costs £3.99.

A slice of a circular 14-inch deep-pan pepperoni pizza costs £5.49.

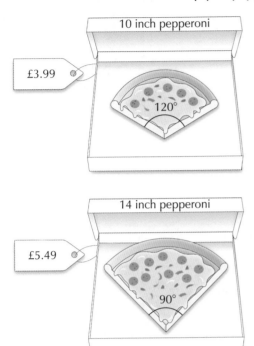

10 inch pepperoni

£3.99

120°

14 inch pepperoni

£5.49

90°

Which of the pizza slices is the better value for money?

You **must** show your working.

_____

_____

_____

_____

_____

_____

_____

_____

_____

_____

_____

(7 marks)

Assessing understanding and problem solving

18 The ratio height : length : depth of this cuboid is 2 : 3 : 4

The total surface area is 1300 cm^2.

Find the volume of the cuboid.

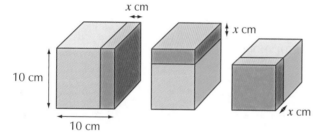

_____

_____

_____

_____

_____

_____

_____

(7 marks)

19 A cube has a side length of 10 cm.

Three slices, each of thickness x cm, are cut off the cube, one after another.

A slice of thickness x cm is removed from the right-hand side.

A slice of thickness x cm is then removed from the top.

A slice of thickness x cm is then removed from the front.

What is the volume of the remaining piece, in terms of x?

_____

_____

_____

_____

_____

_____

(6 marks)

20 A glass paperweight is in the shape of a cone.

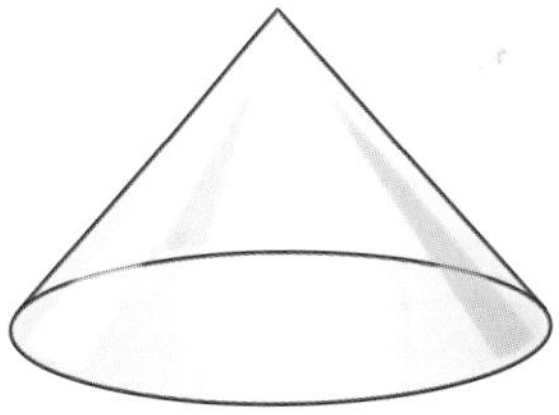

The density of the glass is $2\,\text{g/cm}^3$.

The slant height of the cone is $10\,\text{cm}$.

The vertical height of the cone is $6\,\text{cm}$.

The total surface area of the cone is $144\pi\,\text{cm}^2$.

Work out the mass of the paperweight, in terms of π.

_____

_____

_____

_____

_____

_____

_____

(5 marks)

21 The diagram shows two blue diagonals drawn on the faces of a cube.

Calculate the angle between the blue diagonals.

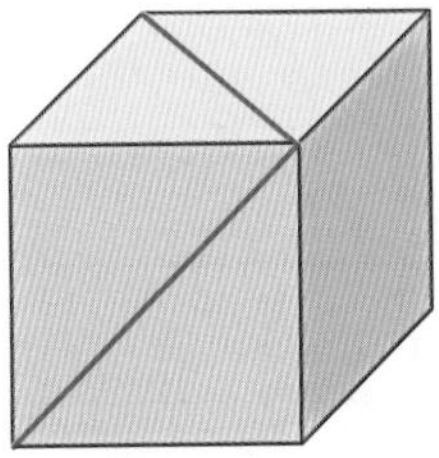

_____

_____

_____

_____

_____

_____

_____

(5 marks)

Assessing understanding and problem solving

22 A rectangle is placed symmetrically inside a square.

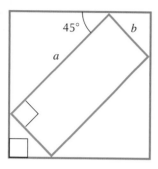

The rectangle has sides of length a and b.

Show that the area of the square is $\frac{1}{2}(a + b)^2$

_____

_____

_____

_____

_____

_____

_____

_____

_____

_____

_____

_____

_____

_____

_____

(10 marks)

23 Sasha is the manager of a shop.

She inspects a delivery of a large batch of china mugs, packaged in boxes of eight.

On this day, 10% of all the china mugs are substandard.

She chooses a box at random and checks for quality.

a Find the probability that the box chosen has no substandard mugs.

_____

_____

(2 marks)

b Find the probability that the box chosen has two or more substandard mugs.

_____

_____

_____

_____

(5 marks)

If the manager finds no substandard mugs she accepts the whole batch.

If she finds two or more substandard mugs she rejects the whole batch.

If she finds one substandard mug she checks a second box of mugs and only accepts the whole batch if there are no substandard mugs in the second box.

c Find the probability that the whole batch is accepted.

_____

(5 marks)

Spot the errors

In each of the questions in this section, look at the two students' answers and mark what they have done wrong.

Some answers have one error and other answers have more than one error. Mark the errors in the book and then work out the correct answer for each question.

Q1 When diesel fuel is sold, the money from the sale is divided in the following ratio.

Tax	:	Oil company	:	Petrol station
77	:	43	:	5

a How many pence in every pound goes to the petrol station? (3 marks)

b One week, a petrol station sells £40 300 worth of diesel fuel.

   i How much of the £40 300 goes to the oil company? (2 marks)

   ii The amount £40 300 was rounded to the nearest £10.

   The cost of diesel fuel is 122 p per litre, to the nearest penny.

   Calculate the maximum number of litres of diesel fuel sold that week. (3 marks)

> **Hint:** This answer has been marked, to show you what to do.

A1

Arnold

a 77 + 43 + 5 = 125p   ✓

125 − 25 = 100p = £1   ✗ $\frac{5}{125}$ × 100 (5p out of 125p → petrol station)

so 25p   so 4p

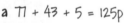 Need all 3 parts of the ratio.

b i $\frac{43}{77 + 5}$ × 40 300 = £21 132.93   ✗ $\frac{43}{125}$ × 40 300 = £13 863.20

ii 40 310 ÷ 1.21 = 33 314 litres   ✗ $\frac{40 305}{1.215}$ = 33 173 litres (nearest litre)

Incorrect bounds used.

A1

Ziggi

a 77 + 43 + 5 = 125

so   5 × $\frac{125}{100}$ = 6.25p

b i $\frac{77 + 5}{125}$ × 40 300 = £26 436.80

ii 40 305 ÷ 121.5 = 331.72 litres

Q2 The total length of all the fish that can safely live in a fish tank can be estimated using this formula:

$$L = \frac{5VF}{10\,000}$$

where L = length of mature medium-sized fish

   V = volume of tank, in cubic centimetres

   F = filter type.

The value of F for different types of filters is shown in the table on the right.

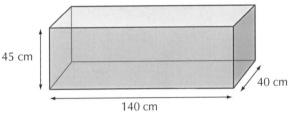

Filter type	F
Under-gravel	1.0
Internal, standard	1.2
Internal, over-sized	1.4
External, standard	1.7
External, over-sized	2.0

Oscar buys a tank with dimensions as shown on the right, below.

Oscar has a list of the mature medium-sized fish that he wants to buy. He estimates that the total length of these fish is 175 cm.

Oscar needs to make sure that these fish can live safely in the tank. What type of filter should he buy?

You **must** show your working.

45 cm

40 cm

140 cm

(4 marks)

A2   Rachel

$L = 1.75\,m$

$V = 0.45 \times 1.40 \times 0.4 = 0.252\,m^3 = 0.252 \times 100 = 25.2\,cm^3$

$1.75 = \dfrac{5 \times 25.2 \times F}{10\,000}$

$17\,500 = 126 \times F$

$F = 138.888$

None of them.

A2   Anil

$L = 175$

$V = 45 + 140 + 40 = 225\,cm^3 = 11\,390\,625$

$175 = \dfrac{5 \times 11\,390\,625 \times F}{10\,000}$

$1\,750\,000 = 56\,953\,125F$

$F = 0.03$

Any of them, I'd go for the cheapest.

Spot the errors

Q3 The difference between the squares of two consecutive even numbers is twice the sum of the numbers.

For example, $10^2 - 8^2 = 36$ and $2 \times (10 + 8) = 36$

Prove this result algebraically.

(5 marks)

A3

Ian

$4^2 - 2^2 = 12$ and $2 \times (4 + 2) = 12$

$6^2 - 4^2 = 20$ and $2 \times (6 + 4) = 20$

$8^2 - 6^2 = 28$ and $2 \times (8 + 6) = 28$

They always go up in 8s so they will always work.

A3

Priya

The first even number is $2n$ so the one below it is $2n - 2$

$2n^2 - 2n^2 - 2 = 2 \times (2n + 2n - 2)$

$\qquad -2 = 8n - 4$

$\qquad\quad 2 = 8n$

$\qquad 8 \div 2 = 4$, which is even.

Q4 O is the centre of the circle.

Angle DCA = 128°

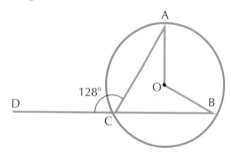

Work out the size of the reflex angle AOB.

You **must** show your working.

(5 marks)

A4

Zac

A = 128° (alternate angle theorem)

A = B = 128° (symmetrical)

C = 180° − 128° = 52° (angles on a straight line)

Angle at O = 360° − 128° − 128° − 52°

= 52° (quadrilateral = 360°)

Reflex angle = 360° − 52°

= 308° (reflex angle bigger than 180°)

A4

Cara

C = 60° (triangle ABC is equilateral)

Reflex angle = 180 − 60

= 120 (opposite angles in cyclic quadrilateral = 180°)

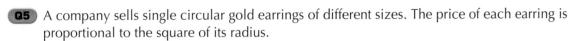

Q5 A company sells single circular gold earrings of different sizes. The price of each earring is proportional to the square of its radius.

The price of a small earring of radius 1.5 cm is £7.50. What is the price of a large earring of radius 3 cm?

(5 marks)

A5

Kristel

The radius of the square earring is double, so

£7.50 × 2 = £15

A5

Tariq

$p \propto r^2$

$p = kr^2$

$1.5 = k \times 7.5^2$

$k = 0.26666$

$p = 0.26666r^2$

$p = 0.26666 \times 3^2$

$p = £24$

Spot the errors

Q6 Study this flowchart.

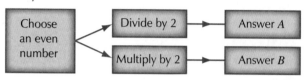

Choose an even number → Divide by 2 → Answer *A*

Choose an even number → Multiply by 2 → Answer *B*

Explain why (*B* – *A*) is always a multiple of 3.

(4 marks)

A6

Timothy

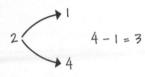

2 → 1, 4 $4 - 1 = 3$

4 → 2, 8 $8 - 2 = 6$

6 → 3, 12 $12 - 3 = 9$

Happens every time.

A6

Laura

n $\frac{1}{2}n$, $2n$ $2n - \frac{1}{2}n = 1\frac{1}{2}n$ which is 3×0.5

$2n$ n, $4n$ $4n - n = 3n$ which is 3×1

$3n$ $1\frac{1}{2}n$, $6n$ $6n - 1\frac{1}{2}n = 4\frac{1}{2}n$ which is 3×1.5

The number is always a multiple of 3.

Q7 *a* and *b* are integers such that $a > 0$, $b > 0$ and:

$$\sqrt{a^2 + 8b} = 11$$

$$\sqrt{a^2 - 16b} = 1$$

Calculate the value of $\sqrt{a^2 - ab}$

(6 marks)

A7

Sanjay

$a + 4b = 11$ ①

$a - 8b = 1$ ②

① − ② gives

$-4b = 10$

$\underline{b = -2.5}$

so $a + -4 \times 2.5 = 11$

$a - 10 = 11$

$\underline{a = 21}$

so $\sqrt{a^2 - ab} = \sqrt{21^2 - 21 \times -2.5} = 13.1339255...$

A7

Charlie

$a^2 + 8b = 121$

$a^2 - 16b = 1$

$24b = 120$

$b = 5$

$a^2 + 8 \times 5 = 121$

$a^2 + 40 = 121$

$a^2 = 81$

$a = 9$

so $\sqrt{a^2 - ab} = \sqrt{9^2 - 9 \times 5} = -36$

Spot the errors

Q8 This right-angled triangle has sides y cm, x cm and $(x + 1)$ cm.

x and y are both integers.

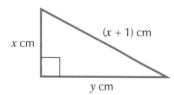

Prove that y must be an odd number.

(4 marks)

A8

Gianni

Pythagorean triples

$y = 3, x = 4, x + 1 = 4 + 1 = 5$ and $3^2 + 4^2 = 5^2$

$y = 5, x = 12, x + 1 = 12 + 1 = 13$ and $5^2 + 12^2 = 13^2$

$y = 7, x = 24, x + 1 = 24 + 1 = 25$ and $7^2 + 12^2 = 13^2$

y is always an odd number because there must be loads more of these.

A8

Cheryl

$x^2 + y^2 = (x + 1)^2$

$x^2 + y^2 = x^2 + 2x + 1$

$y^2 = 2x + 1$, which is the same as $2n + 1$, which is odd, so y must be odd.

Q9 Shami is using the quadratic formula to solve a quadratic equation.

After correctly substituting the values, she writes:

$$x = \frac{8 \pm \sqrt{64 - 96}}{6}$$

a What is the quadratic equation Shami is trying to solve?

(3 marks)

b Explain why Shami will **not** be able to find any solutions to the equation.

(1 mark)

A9

Alicia

a $a = 3$ (because $2 \times 3 = 6$)

$b = 8$ (because $82 = 64$)

$c = -8$ (because $4 \times 3 \times -8 = -96$)

The equation is $3x^2 + 8x - 8$

b She can – it's $\dfrac{8 \pm \sqrt{32}}{6}$

$x = \dfrac{8 + \sqrt{32}}{6}$ or $x = \dfrac{8 - \sqrt{32}}{6}$

$x = 8.94$ or $x = 7.05$ (to 2dp)

A9

Jake

a You can't tell because you cannot find the square root of -32.

b Shami can if she has a calculator!

$x = \dfrac{8 \pm (8 - 9.7979)}{6}$

$x = \dfrac{8 \pm 1.7979}{6}$

$x = 1.03$ or 1.63 (to 2dp)

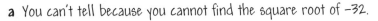

Q10 Expand and simplify $(4 + \sqrt{5})^2$

(3 marks)

A10

Lola

$(4 + \sqrt{5})^2 = (4 + \sqrt{5})(4 + \sqrt{5})$

$= 16 + 5$

$= 21$

A10

Ciaran

$(4 + \sqrt{5})^2 = 8 + 5\sqrt{4} + 5\sqrt{4} + 5$

$= 13 + 10\sqrt{4}$

$= 13 + 10 \times 2$

$= 33$

Spot the errors

Q11 Prove that $\dfrac{\sqrt{20} + 10}{\sqrt{5}} = 2(1 + \sqrt{5})$

(4 marks)

A11

Leo

Left-hand side	Right-hand side
$= \dfrac{\sqrt{20}}{\sqrt{5}} + 10$	$= 2 + \sqrt{5}^2$
$= \sqrt{4} + 10$	$= 2 + 10$
$= 2 + 10$	$= 12$
$= 12$	

Both sides = 12, proved

A11

Roxy

$\dfrac{\sqrt{20} + 10}{\sqrt{5}} = 2(1 + \sqrt{5})$ ÷ by 2

$\dfrac{\sqrt{10} + 5}{\sqrt{5}} = (1 + \sqrt{5})$ × by $\sqrt{5}$

$\sqrt{10} + 5 = \sqrt{5} + 2\sqrt{5}$

$\sqrt{10} + 5 = 3\sqrt{5}$

$2\sqrt{5} + 5 = 3\sqrt{5}$

$3\sqrt{5} = 3\sqrt{5}$

LHS = RHS, proved

Q12 Harriet's scores on a computer game are summarised in the table on the right.

Draw a histogram for this data.

(6 marks)

Score, s	Frequency
$0 \leqslant s < 20$	10
$20 \leqslant s < 40$	18
$40 \leqslant s < 50$	25
$50 \leqslant s < 60$	20
$60 \leqslant s < 80$	10
$80 \leqslant s < 100$	5

SPOT THE ERRORS

A12 Rico

Harriet's computer game scores

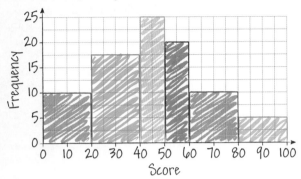

A12 Nate

Score, s	Frequency	Frequency density
$0 \leq s < 20$	10	2
$20 \leq s < 40$	18	1.1
$40 \leq s < 50$	25	0.4
$50 \leq s < 60$	20	0.5
$60 \leq s < 80$	10	2
$80 \leq s < 100$	5	4

Histogram of Harriet's computer game scores

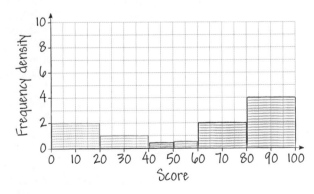

Spot the errors

Q13 A bag contains two red counters and eight blue counters.

A counter is drawn from the bag at random and not replaced; then a second counter is drawn.

Use a tree diagram to find the probability that the second counter drawn is red.

(5 marks)

A13 Sophie

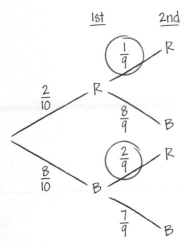

Probability that 2nd is red = $\frac{1}{9} + \frac{2}{9} = \frac{3}{9}$

A13 Asim

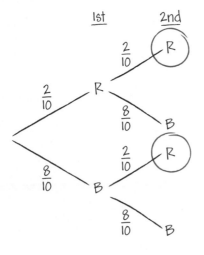

Prob 2nd is red = $\frac{2}{10} \times \frac{2}{10} + \frac{8}{10} \times \frac{8}{10} = \frac{4}{100} + \frac{64}{100} = \frac{68}{100}$

Number

Moving from B to A

A

Complete calculations involving fractional and negative numbers and powers (A)

Able to write surds in their simplest form (A)

Solve complex problems involving percentage increases and decreases (A)

Complete compound interest and depreciation calculations by formula (A)

Able to interpret numbers with negative or simple fractional powers (B)

Complete calculations involving standard index form (B)

Solve problems involving HCF (Highest Common Factor) or LCM (Lowest Common Multiple) in context (B)

Solve problems involving reverse percentages (B)

Able to construct and use formulae for direct and inverse variation to solve problems (A)

Solve simple problems involving direct proportional change (B)

Able to convert any recurring decimal to a fraction in its simplest form (A)

Know how to convert recurring decimals to fractions (A)

Able to approximate to given or appropriate degrees of accuracy (B)

Know how to determine an appropriate degree of accuracy for an answer from its context (B)

SKILLS **B** CONTENT

Know how to interpret, order and calculate with numbers written in standard index form (B)

Know how to construct and use formulae for direct and inverse variation (A)

Know how to complete a reverse percentage calculation (B)

Know how to write and interpret numbers written with negative and simple fractional powers (B)

Find measures of accuracy for numbers given to decimal places or significant figure accuracies (A)

Know how to complete compound interest and depreciation calculations by using the appropriate formula (A)

Know how to use the rules of indices for negative and fractional numbers and powers (A)

Recognise that some square roots are irrational and should be written as surds (A)

Know how to simplify surds (A)

A

Algebra

Moving from B to A

Progression map

A

Know how to identify and factorise the difference of two squares (A)

Know how to factorise a quadratic expression of the form $ax^2 + bx + c$ (A)

Know how to solve a quadratic equation of the form $ax^2 + bx + c = 0$ by factorisation (A)

Know how to use the quadratic formula to solve a quadratic equation of the form $ax^2 + bx + c = 0$ (A)

Know how to construct (and solve) simultaneous equations from a practical situation (A)

Understand the identity symbol and how and when it is used (A)

Know how to rearrange a formula where the subject appears twice (A)

Know how to factorise a simple quadratic expression (B)

Know how to factorise and solve a simple quadratic equation, understanding why there may be two solutions (B)

Know how to solve a pair of simple linear simultaneous equations algebraically (B)

Know how to solve a pair of linear simultaneous equations from their graphs (B)

Know how to combine algebraic fractions using $+, -, \times, \div$ (A)

Know how to represent inequalities in one or two variables graphically (B)

Know how to multiply out a pair of linear brackets (B)

Understand that for a pair of simultaneous equations, there will be a unique pair of values that will satisfy both equations (B)

Understand gradients in parallel lines (B)

Know how to plot more complex graphs (such as cubics) using a table of values (B)

Recognise the shapes of the graphs $y = x^3$ and $y = \frac{1}{x}$ (B)

B CONTENT

SKILLS B

Can represent a region that satisfies one linear inequality, or more, graphically (B)

Can represent a region that satisfies one linear inequality, or more, graphically (B)

Can expand and simplify a pair of linear brackets to get a quadratic expression (B)

Able to factorise a quadratic expression of the form $x^2 + bx + c$ (B)

Can draw straight line graphs using the gradient-intercept method (B)

Able to derive graphs from real life situations (linear and simple quadratic) (B)

Able to interpret real-life graphs (B)

Able to manipulate and simplify algebraic fractions (A)

Able to solve linear inequalities in two variables (B)

Able to solve a quadratic equation of the form $x^2 + bx + c = 0$ (B)

Can rearrange more complicated formulae (B)

Can solve two simple linear simultaneous equations algebraically (B)

Able to solve a pair of linear simultaneous equations from their graphs – possibly in context (B)

Can draw a variety of graphs such as exponential and reciprocal, using a table of values (A)

Can interpret and draw more complex real-life graphs and use them to solve problems (A)

Able to use the quadratic formula accurately to solve a quadratic equation of the form $ax^2 + bx + c = 0$ (A)

Able to solve a quadratic equation of the form $ax^2 + bx + c = 0$ by factorisation (A)

Able to factorise a quadratic expression of the form $ax^2 + bx + c$ (A)

Can accurately rearrange a formula where the subject appears twice (A)

Can set up and solve two simultaneous equations from a practical problem (A)

A

Geometry and Measures

Moving from B to A

A

A

Use the sine and cosine rules to calculate missing angles or sides in non-right-angled triangles, in 2D (A)

Use the formula $A = \frac{1}{2} ab \sin C$ to solve problems involving areas of triangles (A)

Solve angle problems in circles using circle theorems (A)

Solve problems using area and volume scale factors (A)

Solve practical problems using similar triangles (A)

Solve problems using addition and subtraction of vectors (A)

Understand that vectors represent movement, and that vectors can be combined (A)

SKILLS B CONTENT

B

Complete compound transformations (B)

Enlarge a 2D shape by a fractional scale factor (B)

Enlarge a 2D shape by a negative scale factor (B)

Construct a perpendicular from a point to a line (B)

Find angles in circles using circle theorems (B)

Construct an angle of 60° (B)

Calculate the surface area and/or volume of pyramids, cones and spheres (B)

Solve problems involving effects of enlargement (B)

Know formulae/ methods to calculate surface areas and volumes of 3D shapes (B)

Know how to set up equations to find missing sides in similar triangles (B)

Use trigonometry to solve problems (B)

Solve problems in 3D using Pythagoras' theorem (B)

Use trigonometry to find lengths of sides and angles in right-angled triangles (B)

Know and describe how transformations can be combined (B)

Understand and use a range of compound measures (B)

Solve more complex 2D (and 3D) problems using Pythagoras' theorem and trigonometry (A)

Know the effect of enlargement on area and volume scale factors (A)

Know and use circle theorems to find angles in circles (B)

Know trigonometric ratios (sin, cos, tan) for right-angled triangles (B)

Know how to calculate the length of an arc and the area of a sector (B)

Know the conditions to show that two triangles are congruent (B)

Know how to prove that two triangles are congruent (A)

Know how to find angles in circles using the alternate segment theorem (A)

Know sine and cosine rules and when to use which in non-right-angled triangles (A)

Know how to find the area of a triangle using the formula $A = \frac{1}{2} ab \sin C$ (A)

Know the effect of enlargement on area and volume scale factors (A)

Progression map

Statistics and Probability

Moving from B to A

Know how to calculate frequency density for grouped data (A)

Know how a histogram differs from a bar chart and how to construct one (A)

Draw histograms from frequency tables with unequal class intervals (A)

Know how to construct a box plot for given data (B)

Know how to construct a cumulative frequency table and diagram (B)

Understand how to use a tree diagram to work out the probability of combined events (A)

Know how to calculate combined probabilities for independent events (B)

SKILLS **B** CONTENT

Complete a cumulative frequency table (B)

Understand why a sample may be needed for a statistical investigation and different methods of obtaining a sample (A)

Draw and use a tree diagram to work out the probability of combined events (B)

Draw a cumulative frequency diagram (B)

Use box plots to compare distributions (B)

Draw box plots for given data (B)

Calculate the numbers to be surveyed for a stratified sample (A)

Solve problems using cumulative frequency diagrams (A)

Find median and quartiles from a cumulative frequency diagram (A)

Compare measures of spread using cumulative frequency diagrams and/or box plots (A)

Use and/or combinations of a tree diagram to work out probabilities of specific outcomes of combined events (A)

Number

Moving from A to A*

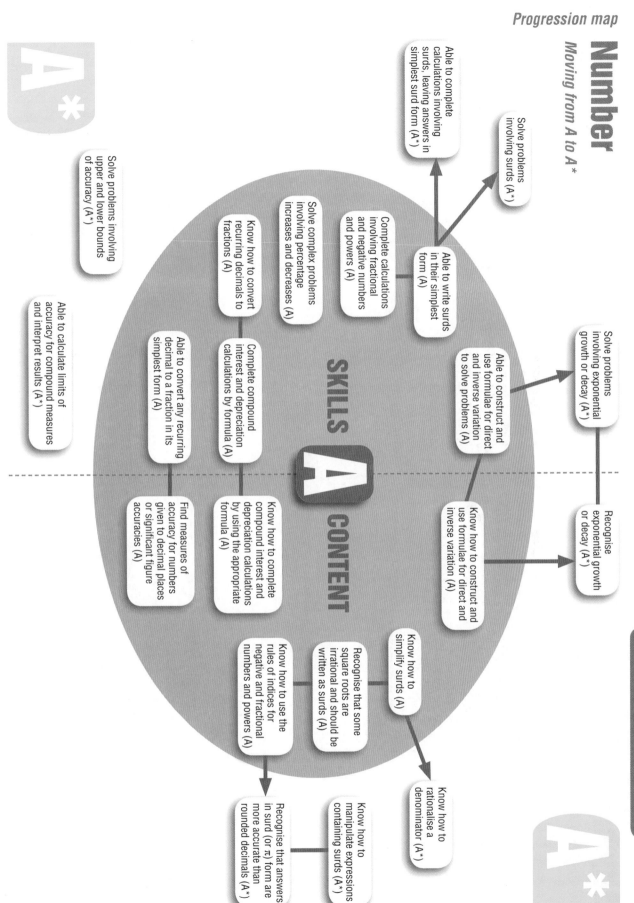

SKILLS A CONTENT

Able to complete calculations involving surds, leaving answers in simplest surd form (A*)

Solve problems involving surds (A*)

Complete calculations involving fractional and negative numbers and powers (A)

Able to write surds in their simplest form (A)

Solve complex problems involving percentage increases and decreases (A)

Know how to convert recurring decimals to fractions (A)

Solve problems involving upper and lower bounds of accuracy (A*)

Able to calculate limits of accuracy for compound measures and interpret results (A*)

Able to convert any recurring decimal to a fraction in its simplest form (A)

Complete compound interest and depreciation calculations by formula (A)

Know how to complete compound interest and depreciation calculations by using the appropriate formula (A)

Find measures of accuracy for numbers given to decimal places or significant figure accuracies (A)

Able to construct and use formulae for direct and inverse variation to solve problems (A)

Know how to construct and use formulae for direct and inverse variation (A)

Solve problems involving exponential growth or decay (A*)

Recognise exponential growth or decay (A*)

Know how to use the rules of indices for negative and fractional numbers and powers (A)

Recognise that some square roots are irrational and should be written as surds (A)

Know how to simplify surds (A)

Know how to rationalise a denominator (A*)

Know how to manipulate expressions containing surds (A*)

Recognise that answers in surd (or π) form are more accurate than rounded decimals (A*)

Algebra

*Moving from A to A**

A*

A

Know how to recognise and solve quadratic equations in context or when embedded in practical problems (A*)

Know how to solve a quadratic equation by completing the square (A*)

Know how to solve a pair of simultaneous equations where one is linear and the other is either quadratic or of the form $x^2 + y^2 = r^2$ (A*)

Know how to use intersections of graphs to solve equations (A*)

Know how to transform graphs of given functions (A*)

Know how to identify and describe transformations of graphs (A*)

Know how to factorise a quadratic expression of the form $ax^2 + bx + c$ (A)

Know how to solve a quadratic equation of the form $ax^2 + bx + c = 0$ by factorisation (A)

Know how to use the quadratic formula to solve a quadratic equation of the form $ax^2 + bx + c = 0$ (A)

Know how to rearrange formulae where the subject appears twice or as a power (A*)

Know how to rearrange a formula where the subject appears twice (A)

Know how to identify and factorise the difference of two squares (A)

Understand the identity symbol and how and when it is used (A)

Know how to construct (and solve) simultaneous equations from a practical situation (A)

Know how to combine algebraic fractions using $+, -, \times, \div$ (A)

Know how to combine and simplify algebraic fractions using factorisation (A*)

Know how to solve equations involving algebraic fractions (A*)

CONTENT

A

SKILLS

Use inequalities to describe practical situations and find possible solutions (A)

Able to solve a quadratic equation of the form $ax^2 + bx + c = 0$ by factorisation (A)

Can accurately rearrange a formula where the subject appears twice (A)

Can draw a variety of graphs such as exponential and reciprocal, using a table of values (A)

Able to manipulate and simplify algebraic fractions (A)

Simplify algebraic fractions by factorisation and cancellation (A*)

Solve a quadratic equation obtained from manipulating algebraic fractions where the variable appears in the denominator (A*)

Able to use the quadratic formula accurately to solve a quadratic equation of the form $ax^2 + bx + c = 0$ (A)

Able to factorise a quadratic expression of the form $ax^2 + bx + c$ (A)

Can set up and solve two simultaneous equations from a practical problem (A)

Can interpret and draw more complex real-life graphs and use to solve problems (A)

Can transform the graph of a given function (A*)

Can identify the equation of a function from its graph, which has been formed by a transformation on a known function (A*)

Able to solve a quadratic equation by completing the square (A*)

Able to solve real-life problems that lead to constructing and solving a quadratic equation (A*)

Can rearrange more complicated formulae where the subject may appear twice or as a power (A*)

Can solve a pair of simultaneous equations where one is linear and the other is either quadratic, or of the form $x^2 + y^2 = r^2$ (A*)

Able to solve equations using the intersections of two graphs (A*)

A*

Geometry and Measures

Moving from A to A*

A*

A

SKILLS A CONTENT

Solve 3D problems using Pythagoras' theorem and trigonometric ratios (A*)

Find the angle between a line and a plane in a 3D shape (A*)

Use the sine and cosine rules to solve more complex problems involving non-right-angled triangles (A*)

Use circle theorems to prove geometrical results (A*)

Use the formula $A = \frac{1}{2} ab \sin C$ to solve problems involving areas of triangles (A)

Use the sine and cosine rules to calculate missing angles or sides in non-right-angled triangles in 2D (A)

Solve more complex 2D (and 3D) problems using Pythagoras' theorem and trigonometry (A)

Solve simple problems using addition and subtraction of vectors (A)

Solve more complex geometrical problems using vectors (A*)

Know how to calculate with vectors in 2D, use commutative and associative laws and find resultants (A*)

Solve related problems to prove geometrical results (A*)

Solve related problems involving, e.g. capacity, using area and volume scale factors (A*)

Solve problems using area and volume scale factors (A)

Solve angle problems in circles using circle theorems (A)

Solve practical problems using similar triangles (A)

Know the effect of enlargement on area and volume scale factors (A)

Know how to find angles in circles using the alternate segment theorem (A)

Understand that vectors represent movement, and that vectors can be combined (A)

Know how to find the area of a triangle using the formula $A = \frac{1}{2} ab \sin C$ (A)

Know how to calculate complex surface areas and volumes, e.g. for the frustum of a cone (A*)

Know how to prove that two triangles are congruent (A)

Know sine and cosine rules and when to use which in non-right-angled triangles (A)

Know how to complete a geometric proof using algebra to show the general case (A)

Know how to identify where and how to use Pythagoras' theorem or trigonometry to solve 3D problems (A*)

A*

Statistics and Probability

Moving from A to A*

SKILLS **A** **CONTENT**

Know how to calculate frequency density for grouped data (A)

Know how a histogram differs from a bar chart and how to construct one (A)

Understand why a sample may be needed for a statistical investigation and different methods of obtaining a sample (A)

Understand how sample size and structure can affect results (A*)

Understand how bias can affect a sample or results (A*)

Draw histograms from frequency tables with unequal class intervals (A)

Solve problems using cumulative frequency diagrams (A)

Find median quartiles from a cumulative frequency diagram (A)

Compare measures of spread using cumulative frequency diagrams and/or box plots (A)

Calculate the numbers to be surveyed for a stratified sample (A)

Use and/or combinations of a tree diagram to work out probabilities of specific outcomes of combined events (A)

Understand how one event can be affected by another in calculating conditional probabilities (A*)

Use tree diagrams to solve problems involving conditional probability (A*)

Solve problems involving histograms and frequency density (A*)

ANSWERS TO NUMBER

Estimating with square roots

1 a 0.2 **b** 0.5 **c** 0.8 **d** 1.1 **e** 1.5
f 0.12 **g** 0.013 **h** 0.009 **i** 0.05
2 a 6 **b** 15 **c** 1.2 **d** 0.3 **e** 0.6
f 20
3 a 60 **b** 5.9 **c** 10 **d** 10
e 0.6 **f** 51 **g** 0.7 or 0.8
h 13 or 13.2 **i** 10

Reverse percentages

1 20 kg **2** 68 kg **3** £2550
4 7.2 million dollars
5 278 mm **6** $7 200 000

Standard index form

1 6.023×10^{23}
2 $1.37 \times 10^{18} \text{m}^3$
3 $2 \times 10^{-8} \text{m}$
4 $3.3 \times 10^{-25} \text{g}$
5 a 10000 **b** 120000 **c** 0.0000123 **d** 0.1234
6 a 3.4×10^4 **b** 2.6×10^4 **c** 1.2×10^8 **d** 7.5×100
7 $1.53 \times 10^{13} \text{cm}$
8 $2 \times 10^0 \text{kg}$
9 135 kg

Limits

1 a 5.65–5.75 kg **b** 5.195–5.205 mg
c 89.9985–89.9994 Gb
2 928.5 kg
3 a 541.5 cm² **b** 1157.625 cm³
4 No, the maximum size of the card is 12.55 cm by 8.5 cm and the minimum size of the envelope is 12.5 cm by 9.5 cm, so the card might be too long.
5 a $45 \div 6.5 = 6.923$ seconds
b $45 \div 1.325 = 33.96 = 34$ steps

Indices

1 a 1 **b** 1 **c** 1
2 a $\frac{1}{8}$ **b** $\frac{1}{125}$ **c** $\frac{1}{10000}$ **d** $\frac{1}{12}$ **e** $\frac{1}{x}$ **f** $\frac{1}{x}$
3 a 3^{-2} **b** t^{-3} **c** h^{-m}
4 a $4^2 = 16$ **b** $4^4 = 256$ **c** 1 **d** $2^{-4} = \frac{1}{16}$
5 a $\frac{2}{t^3}$ **b** $\frac{3}{4^p}$
6 a 5 **b** 5 **c** $\frac{1}{4}$ **d** 2 **e** 10 **f** $\frac{1}{2}$
7 a 4 **b** 16 **c** 100 **d** 32 **e** $\frac{1}{9}$ **f** $\frac{1}{10^5}$

Recurring decimals to fractions

1 a $\frac{1}{9}$ **b** $\frac{2}{9}$ **c** $\frac{3}{9} = \frac{1}{3}$ **d** $\frac{4}{9}$ **e** $\frac{7}{9}$ **f** $\frac{8}{9}$
2 If $x = 0.\dot{9}$, then $10x = 9.\dot{9}$, $9x = \frac{9}{9}$ and $\frac{9}{9} = 1$. So, $0.\dot{9} = 1$
3 a $\frac{23}{99}$ **b** $\frac{1}{45}$ **c** $\frac{73}{300}$ **d** $\frac{367}{1665}$

Surds

1 a $\sqrt{15}$ **b** $2\sqrt{15}$
2 a 5 **b** 6 **c** 20 **d** 12 **e** 10 **f** 3
3 a $2\sqrt{3}$ **b** $4\sqrt{5}$
4 a $6\sqrt{15}$ **b** $18\sqrt{6}$ **c** $4\sqrt{5}$ **d** 10 **e** $13\sqrt{2}$ **f** $3\sqrt{3}$
5 a 12 **b** 20
6 a $4\sqrt{3}$ **b** $3\sqrt{2}$ **c** $\frac{1}{2}$ **d** $2(\sqrt{3} - 1)$ or $2\sqrt{3} - 2$
7 a $11 + 6\sqrt{3}$ **b** $\sqrt{3} - 16$ **c** $29 - 4\sqrt{7}$
8 $(\sqrt{15} - \sqrt{12})(\sqrt{15} + \sqrt{12}) =$
$15 + \sqrt{15}\,\sqrt{12} - \sqrt{12}\,\sqrt{15} - 12 = 3$
9 $\frac{16 + \sqrt{3}}{12}$
10 a $7^2 + (\sqrt{2} + \sqrt{32})^2 = 49 + 2 + 2\sqrt{64} + 32 = 83 + 16$
$= 99$ and $(3\sqrt{11})^2 = 99$, which satisfies Pythagoras' theorem, so the triangle is right-angled.
b $\frac{1}{2} \times 7(\sqrt{2} + \sqrt{32}) = \frac{1}{2} \times 7(\sqrt{2} + 4\sqrt{2}) = \frac{1}{2} \times 7 \times 5\sqrt{2}$
$= \frac{35\sqrt{2}}{2}$

ANSWERS TO ALGEBRA

Solving quadratic equations graphically

1 a $x = -3$, $x = 1$
b $x = -3.2$ to -3.4, $x = 1.2$ to 1.4
2 a $y = x^2 - 7x + 5$ **b** $x = 0.8$, $x = 6.2$
3 Line is $y = x + 1$, solution $x = 0.3$, $x = 3.7$

Recognising shapes of graphs

1 $y = x^2 + 4$ is graph G **2** $y = 2x^2 + 4$ is graph I
3 $y = x^2 - 4$ is graph B **4** $y = x^2 + 2x$ is graph E
5 $y = 2x + 4$ is graph A **6** $y = x^3 + 4$ is graph H
7 $y = -x^3 + 4$ is graph F **8** $y = x^2 + 2x + 4$ is graph C
9 $y = \frac{4}{x}$ is graph D

Real-life graphs

1 a 6 m/s² **b** 7.5 m/s² **c** 285 m
2 a 50 km/h² **b** 60 km/h² **c** 35 km/h²
d 75 km **e** 120 km

Answers

3 a

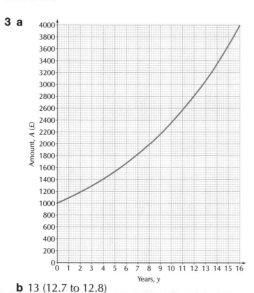

b 13 (12.7 to 12.8)

Equation of a straight line: $y = mx + c$

1 a 4   **b** −0.8

2 a $y = 3x − 1$ **b** $y = −2x + 5$

3 a $y = 2x − 1$ **b** $y = \frac{1}{2}x − 3$

4 $x + 3y = 6$ or $y = −\frac{1}{3}x − 3$ or $y = −\frac{x}{3} + 2$

5 $x + 2y = 8$ or $y = −\frac{1}{2}x + 4$

Solving simultaneous equations graphically

1 $x = 1$, $y = 6$ **2** $x = 1$, $y = 4$

3 $x = −1$, $y = 3$ and $x = 4$, $y = 8$

4 $x = 2$, $y = 8$ and $x = −1$, $y = 2$

Inequalities

1 a $y > 2$ **b** $x > 2$ **c** $y \geqslant x$ **d** $y \leqslant x + 2$

2 a   **b**   **c**

3 $x < 1.5$ $y \leqslant 2$ $y \geqslant x − 1$

4

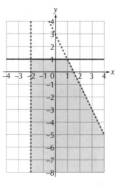

5 a

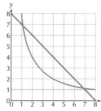

b 2 and 5, 3 and 3, 3 and 4

Drawing complex graphs

1 a

x	−2.5	−2	−1	0	1	1.2
x^3	−15.625	−8	−1	0	1	1.728
$+2x^2$	12.5	8	2	0	2	2.88
−1	−1	−1	−1	−1	−1	−1
$y = x^3 + 2x^2 − 1$	−4.125	−1	0	−1	2	3.608

b

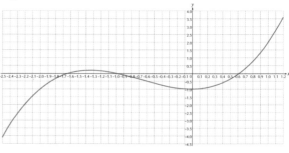

c $x = −1.6$, −1.0, 0.6

2 a

x	−6	−5	−4	−3	−2	−1	1	2	3	4	5	6
$f(x)$	0	−0.4	−1	−2	−4	−10	14	8	6	5	4.4	4

b

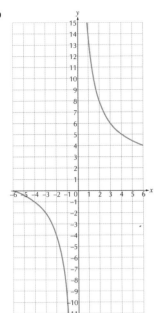

c $x = 0$, $y = 2$

d 5.75

Trigonometric graphs

1 a 36°, 144°　**b** 216°, 324°　**c** 54°, 306°
　d 126°, 234°　**e** 88°, 92°　**f** 88°, 272°
2 a 151°　　　**b** 209°, 331°　**c** 61°, 299°

Transformation of functions

1 a f(x) + 1 = A　**b** f(x) − 1 = B　**c** f(x) − 2 = C
2 a f(x) − 5: A(−10, −5), B(1, −5), C(5, −8)
　b 2 + f(x): A(−10, 2), B(1. 2), C(5, −1)

3 a

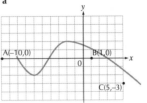

b　　　　　　　　　**c**

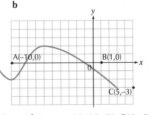

4 a 2f(x):　　A(−10, 0), B(1, 0), C(5, −6)
　b −3f(x):　A(−10, 0), B(1, 0), C(5, 9)
　c $\frac{1}{2}$f(x):　A(−10, 0), B(1, 0), C(5, −$\frac{3}{2}$)
　d 2f(x − 2): A(−8, 0), B(3, 0), C(7, −6)
　e 2f(x) + 1: A(−10, 1), B(1, 1), C(5, −5)
5 a f(2x):　　A(−5, 0), B(½, 0), C(2.5, −3)
　b f(−x):　　A(10, 0), B(−1, 0), C(−5, −3)
　c f($\frac{1}{2}x$):　　A(−20, 0), B(2, 0), C(10, −3)

6

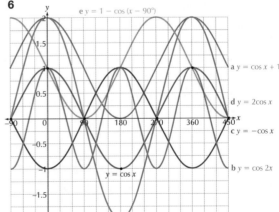

e $y = 1 - \cos(x - 90°)$
a $y = \cos x + 1$
d $y = 2\cos x$
c $y = -\cos x$
b $y = \cos 2x$
$y = \cos x$

3 a

t	0	1	2	3	4	5	6	8	10	12
N	200	400	800	1600	3200	6400	12 800	51 200	204 800	819 200

b

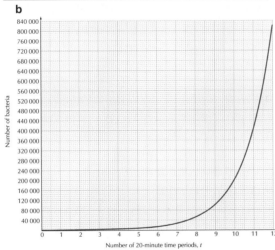

c i About 20 000　**ii** 140 000 to 150 000
d i 180 minutes or 3 hours　**ii** 210 minutes or 3.5 hours

Parallel and perpendicular graphs

1 Parallel to $y = 2x - 2$: **a** Yes　**b** No　**c** No
　　　　　　　　　　d Yes　**e** No　**f** Yes
2 Perpendicular to $y = 2x - 2$: **c** and **d**
3 a $y = -3x \pm$ any number　**b** $y = \frac{1}{3}x \pm$ any number
4 a $y = -3x + 6$　　　**b** $y = \frac{1}{3}x + 9\frac{1}{3}$ or $3y = x + 28$
5 $y = -\frac{1}{2}x + 9$ or $2y = -x + 18$

Answers

Changing the subject of a formula

1 a $r = t^2 - p$ **b** $r = \frac{1}{3}(f^2 + e)$ **c** $r = \left(\frac{d-w}{5}\right)^2$

   d $r = \sqrt{\frac{3V}{\pi h}}$ **e** $r = \sqrt{\frac{A}{\pi} + s^2}$ **f** $r = \frac{gT^2}{4\pi^2}$ or $g\left(\frac{T}{2\pi}\right)^2$

2 a $x = \frac{7-y}{2t}$ **b** $x = \frac{4y+3}{2}$ **c** $x = \frac{bt-ar}{a-b}$

3 a $x = \frac{a}{2b^2 - 3}$ **b** $r = \sqrt{\frac{t-b}{a+c}}$ **c** $x = \frac{acd}{b+ac}$

Solving simultaneous equations: both linear

1 a $x = 6, y = 3$ **b** $x = 2, y = 2.75$ **c** $x = 1, y = 6$

2 a $x = 6, y = 1$ **b** $x = 3, y = -1$ **c** $x = 2.5, y = 1.5$

3 185 sandwiches, 315 baguettes

Solving simultaneous equations: one linear, one non-linear

1 a $x = -1, y = -1$ and $x = -3, y = -5$)
   b $x = 4, y = 8$ and $x = -2, y = 20$

2 a $x = 2, y = -5$) and $x = 5, y = -2$)
   b $x = 2, y = 3$ and $x = -3, y = -2$

3 a $x = -\frac{4}{3}, y = -2$
   b Sketch 2 is correct, as the line touches the curve just once.

Factorising quadratic expressions

1 $(x+2)(x+5)$ **2** $(x+3)(x+7)$ **3** $(x+2)(x+6)$

4 $(x+5)^2$ **5** $(x-2)(x+3)$ **6** $(x-3)(x+5)$

7 $(x-2)(x+6)$ **8** $(x-2)(x+8)$ **9** $(x-3)(x-4)$

10 $(x-3)(x-5)$ **11** $(x+5)(x-6)$ **12** $(x+2)(x-9)$

Solving quadratic equations

1 $(x-2)(x+3) = 0; x = 2, x = -3$

2 $(x-3)(x-5) = 0; x = 3, x = 5$

3 $(x+2)(x+3) = 0; x = -2, x = -3$

4 $(x+3)(x+5) = 0; x = -3, x = -5$

5 $(x-2)(x+7) = 0; x = 2, x = -7$

6 $(x+6)(x-12) = 0; x = -6, x = 12$

7 3

8 −5

Factorising harder quadratic expressions

1 $(3x+2)(x+1)$ **2** $(6x+1)(x+4)$ **3** $(3x-2)(x+4)$

4 $(3x-2)(x+6)$ **5** $(3x-2)(2x-3)$ **6** $(3x-5)(x-2)$

7 $(2x-3)(2x+1)$ **8** $(3x+2)(x-2)$

Solving harder quadratic equations: factorising

1 $(2x+1)(x+3) = 0; x = -\frac{1}{2}, x = -3$

2 $(3x+2)(2x-1) = 0; x = -\frac{2}{3}, x = \frac{1}{2}$

3 $(5x+3)(2x-1) = 0; x = -\frac{3}{5}, x = \frac{1}{2}$

4 $(4x-1)(x-7) = 0; x = \frac{1}{4}, x = 7$

5 a $3x(5x-24)$
   b $15x^2 - 72x - 15 = 0$, so $5x^2 - 24x - 5 = 0$,
     $(5x+1)(x-5) = 0; x = -\frac{1}{5}, x = 5$,
     take the positive value, $x = 5$

Solving quadratic equations: using $x = \frac{-b \pm \sqrt{b^2 - 4ac}}{2a}$

1 $x = -0.27, x = -3.73$ **2** $x = 3.56, x = -0.56$

3 $x = -0.10, x = -9.90$ **4** $x = 1.16, x = -1.56$

5 $x = 2.10, x = -1.43$ **6** $x = 2.58, x = -0.58$

Completing the square

1 a $(x+6)^2 - 29$ **b** $(x+2)^2 - 10$

   c $(x-4)^2 - 10$ **d** $(x+10)^2 - 101$

   e $(x+3\frac{1}{2})^2 - 6\frac{1}{4}$ **f** $(x+2\frac{1}{2})^2 - 6\frac{3}{4}$

2 a $x = -1, x = -13$ **b** $x = 2, x = -4$

   c $x = 5.73, x = 2.27$ **d** $x = 1.70, x = -4.70$

Solving equations with algebraic fractions

1 $x = -5$ **2** $x = 5$ **3** $x = 5.6$ **4** $x = -37$

5 $x = -11\frac{1}{6}$ **6** $x = 46$

Solving harder equations with algebraic fractions

1 $x = 0, x = -2\frac{1}{2}$ **2** $x = -2, x = -3\frac{3}{5}$ **3** $x = 4\frac{3}{4}, x = -1$

4 $x = 5, x = -5$ **5** $x = 2, x = -1\frac{1}{2}$ **6** $x = 3 \pm \sqrt{3}$

Solving linear inequalities

1 a $x < 7$ **b** $x \geqslant 7$ **c** $x \leqslant -70$

   d $x > -3$ **e** $x \geqslant -2$ **f** $x \geqslant 1.5$

2 a $5(x+3) > 2[5 + (x+3)]$ so $5x + 15 > 2(8 + x)$,
     $5x + 15 > 16 + 2x, 3x > 1, x > \frac{1}{3}$

b 1

Simplifying algebraic fractions

1 a $5a$ **b** $5a$ **c** $\frac{1}{2}$ **d** $\frac{2b}{5a}$ **e** $2x + 4$

   f $\frac{2}{3}$ **g** 4   **h** $\frac{2+b}{a}$ **i** $\frac{1}{f-5}$ **j** $a+2$

2 a $x + 7$ **b** $2x$ **c** $\frac{x-4}{x+3}$ **d** $\frac{1}{x+4}$ **e** $\frac{x-2}{3x-2}$

Simplifying algebraic fractions (addition and subtraction)

1 a $\frac{29a}{15}$ **b** $\frac{b-43}{20}$ **c** $\frac{2c+13}{5}$ **d** $\frac{3d+7e}{de}$

2 a $\frac{15a+8}{6a^2}$ **b** $\frac{3(2b-3)}{(b+1)(b-4)}$ **c** $\frac{1-x}{(x-3)(x-5)}$ **d** $\frac{10x-48}{(x-5)^2}$

3 $\frac{2(8x-11)}{(x+3)(x-4)}$ or $\frac{16x-22}{(x+3)(x-4)}$

Simplifying algebraic fractions (multiplication and division)

1 a $\frac{x^2}{15}$ **b** $\frac{5x}{3y}$ **c** $\frac{y}{5}$ **d** $\frac{5}{y}$ **e** $5a$ **f** $5a$

2 a $\frac{(x+2)(x-4)}{15}$ or $\frac{x^2-2x-8}{15}$ **b** $\frac{5(x-3)}{3(x+5)}$

   c $\frac{(t-4)}{4b}$ **d** $\frac{(x+3)}{x^2+3}$

3 a $\frac{3(x+4)}{x-2}$ **b** $\frac{x(x-4)}{(3x-4)^2}$

Proportionality

1 a $x=4y$ **b** 12   **c** 11

2 a $x=\frac{100}{y}$ **b** 10   **c** 4

3 $153.86\,\text{cm}^2$

4 a 10 minutes 40 seconds or $10\frac{2}{3}$ minutes
   b 7 minutes

5 1.25 cm

6 a Table B   **b** Table A   **c** Table C

7 a Sketch 2   **b** Sketch 3   **c** Sketch 1

ANSWERS TO GEOMETRY AND MEASURES

Arcs and sectors

1 a 5.24 cm **b** 9.77 cm   **c** 15.7 cm

2 a 8π **b** $\frac{63}{2}\pi$ or 31.5π **c** $\frac{216}{125}\pi$ or 1.728π

3 a $11.49\,\text{m}^2$ **b** 26

Surface area of cylinders, cones and spheres

1 $301.6\,\text{cm}^2$ **2** $468.2\,\text{cm}^2$

3 $40.5\,\text{m}^2$ **4** $1520.5\,\text{mm}^2$

5 $1592.8\,\text{cm}^2$

Density

1 750 g   **2** $0.038\,\text{g/cm}^3$ **3** $1.6\,\text{g/cm}^3$

4 a 2610 kg   **b** $3480\,\text{kg/m}^3$ **c** 870 kg

5 460 000 tonnes

6 30 cm

7 17.96 or 18.0 m

Volume of cones and spheres

1 a $128\pi\,\text{cm}^3$ **b** $\frac{500}{3}\pi\,\text{cm}^3$ or $166\frac{2}{3}\pi$

2 $11\,600\,\text{cm}^3$ **3** $170\,\text{cm}^3$ **4** 1699 g

Volume and surface area

1 550π **2** 275π

3 a 25 525 kg
   b £3074 (incl. base)   £2493 (excl. base)

Similar shapes

1 No, because $25:20 \neq 19:14$

2 a $x=8.4\,\text{cm}$ **b** $y=7.5\,\text{cm}$ **c** $z=92°$

3 a ∠ABC = ∠EDC (alternate angles), ∠BAC = ∠DEC (alternate angles), ∠ACB = ∠ECD (vertically opposite angles)

b i $x=27\,\text{cm}$ **ii** $y=14.5\,\text{cm}$

4 31.5 m

Scale factors

1 a 3   **b** 9   **c** 27

2 128 square inches   **3** $1200\,\text{cm}^2$

4 6.1 cm

5 a $123.75\,\text{cm}^2$ **b** 533.33 g

6 $94\,770\,\text{cm}^3$

2D trigonometry

1 26.1°   **2** 196 m   **3** 26.03°

4 61.9°   **5** 22.9 km   **6** $153.2\,\text{cm}^2$

7 a 3039 m **b** 547 km/h

3D Pythagoras' theorem and trigonometry problems

1 a 14.1 cm **b** 17.3 cm   **c** 35.3°

2 a 13.3 m **b** 13.0°

3 a 13.8 cm **b** 18.4°   **c** 12.6°

4 67.0°

Sine rule and cosine rule

1 a 4.58 cm **b** 82.1°

2 20.9 km

3 74.8 m

4 17 cm

5 a 132.4 km   **b** 314°

Areas of triangles

1 $2650\,\text{m}^2$ **2** $84\,\text{m}^2$ **3** $10\,708\,\text{m}^2$

Answers

Congruent triangles

1 AB = CD (opposite sides of a parallelogram), BC = DA (opposite sides of a parallelogram), AC is common, so triangle ABC is congruent to triangle CDA (SSS)

Or: ∠BAC = ∠ACD (alternate angles), ∠BCA = ∠CAD (alternate angles), AC is common, so triangle ABC is congruent to triangle CDA (ASA)

2 AB = DE (given), ∠ABC = ∠EDC (alternate angles), ∠BAC = ∠DEC (alternate angles), so triangle ABC is congruent to triangle EDC (ASA)

3 AB = AD (given), BC = DC (given), AC is common, so triangle ABC is congruent to triangle ADC (SSS), therefore ∠ABC = ∠ADC

4 AB = CD (sides of a regular pentagon), BC = DE (sides of a regular pentagon), ∠ABC = ∠CDE (interior angles of a regular pentagon), so triangle ABC is congruent to triangle CDE (SAS)

Circle theorems

1 a $a = 78°$ (angle at the centre is twice the angle at the circumference, when subtended by the same arc)

b $b = 66°$ (angle at the centre is twice the angle at the circumference, when subtended by the same arc)

c $c = 41°$ (angles at the circumference are equal, when subtended by the same arc)

d $d = 67°$ (angles at the circumference are equal, when subtended by the same arc)

e $e = 90°$ (angle in a semicircle)

f $f = 100°$ (sum of opposite angles of a cyclic quadrilateral is 180°)

g $g = 20°$ (obtuse angle at O = 160°, radii meet the tangents at 90°, angle sum of quadrilateral is 360°)

h $h = 50°$ (angle in the alternate segment)

i $i = 10°$ (angle in the alternate segment gives two angles of 85°, angle sum of triangle is 180°)

j $j = 72°$ (supplementary angle to 115° is 65°, third angle in triangle is 72°, which is equal to j (as opposite angles of cyclic quadrilateral, supplementary angles)

k $k = 58°$ (angle in the alternate segment), $l = 58°$ (base angles of isosceles triangle), $m = 76°$ (angle in the alternate segment), $n = 104°$ (opposite angles of cyclic quadrilateral)

2 Let the point where the diameter from B meets the circle be D.
Then ∠BAO = x, ∠AOB = $180° – 2x$ (angle sum of triangle), ∠AOD = $2x$ (angles on a straight line). Similarly, ∠COD = $2y$. Therefore ∠AOC = 2 ∠ABC

3 Join the two points at the bottom to the centre, then use the fact that the angle at the circumference is half the angle at the centre, for both angles.

4 Given a diameter, the angle at the centre is 180° since it is a straight line.
The angle at the circumference is half the angle at the centre.
So the angle in the semicircle is 90°.

Enlargement

1

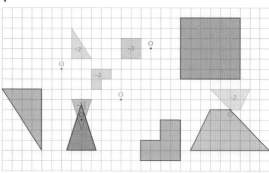

2

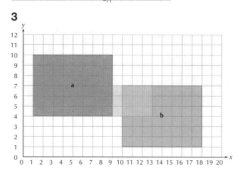

3

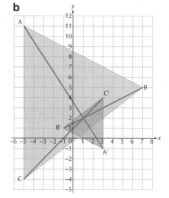

4 a $-\dfrac{4}{3}$

b

c (1, 2)

d Enlargement, scale factor 3, centre (1, 2)

Vectors

1 a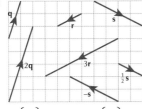

b i $\begin{pmatrix} -3 \\ -9 \end{pmatrix}$ **ii** $\begin{pmatrix} -6 \\ -3 \end{pmatrix}$ **iii** $\begin{pmatrix} 6 \\ -3 \end{pmatrix}$

2 a

b i $\begin{pmatrix} 2 \\ -3 \end{pmatrix}$ **ii** $\begin{pmatrix} -3 \\ 5 \end{pmatrix}$ **iii** $\begin{pmatrix} 6 \\ 4 \end{pmatrix}$

3 a Any two of the single horizontal line segments, e.g. $\overrightarrow{DE}$, $\overrightarrow{JK}$

b Any two of the single sloping line segments, e.g. $\overrightarrow{DH}$, $\overrightarrow{NS}$

c Any two of the single horizontal line segments reversed, e.g. $\overrightarrow{ED}$, $\overrightarrow{KJ}$

d Any two of the single sloping line segments reversed, e.g. $\overrightarrow{HD}$, $\overrightarrow{SN}$

e $\overrightarrow{OC} = 3\mathbf{a}$ **f** $\overrightarrow{FH} = \mathbf{b} - 2\mathbf{a}$ **g** $\overrightarrow{OT} = 3\mathbf{a} + 4\mathbf{b}$

h $\overrightarrow{AN} = \mathbf{a} + 3\mathbf{b}$ **i** $\overrightarrow{IK} = 2\mathbf{a}$ **j** $\overrightarrow{NC} = \mathbf{a} - 3\mathbf{b}$

4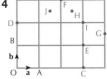

a $\overrightarrow{OC} = 3\mathbf{a}$ **b** $\overrightarrow{OD} = 2\mathbf{b}$

c $\overrightarrow{OE} = 3\mathbf{a} + \mathbf{b}$ **d** $\overrightarrow{OF} = 2\mathbf{a} + 3\mathbf{b}$

e $\overrightarrow{OG} = 4\mathbf{a} + \frac{3}{2}\mathbf{b}$ **f** $\overrightarrow{OH} = \frac{5}{2}(\mathbf{a} + \mathbf{b})$

g $\overrightarrow{IO} = -3\mathbf{a} - 2\mathbf{b}$ **h** $\overrightarrow{JO} = -\frac{3}{2}\mathbf{a} - \frac{5}{2}\mathbf{b}$

5 a $\overrightarrow{AB} = \mathbf{b} - \mathbf{a}$ **b** $\overrightarrow{AM} = \frac{1}{2}(\mathbf{b} - \mathbf{a})$

c $\overrightarrow{BA} = \mathbf{a} - \mathbf{b}$ **e** $\overrightarrow{BM} = \frac{1}{2}(\mathbf{a} - \mathbf{b})$

f $\overrightarrow{OM} = \frac{1}{2}(\mathbf{a} + \mathbf{b})$ **g** $\overrightarrow{MO} = -\frac{1}{2}(\mathbf{a} + \mathbf{b})$

6 a $\overrightarrow{AB} = 2(\mathbf{b} - \mathbf{a})$ **b** $\overrightarrow{DC} = 2\mathbf{b}$

c $\overrightarrow{FE} = -2\mathbf{a}$ **d** $\overrightarrow{EM} = 4\mathbf{b} - \mathbf{a}$

ANSWERS TO STATISTICS AND PROBABILITY

Cumulative frequency diagrams

1 a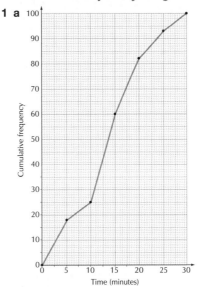

b i 13.6 minutes **ii** 10 minutes

iii 18.4 minutes **iv** 8.5 minutes

c 39 **d** 86 **e** 38

2 a 7g **b** 29 **c** 11.25%

Box plots

1 a i 17 **ii** 24 **iii** 11

b i 210 **ii** 200 **iii** 70

2 a

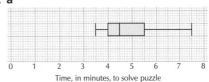

Time, in minutes, to solve puzzle

b Examples of comparisons:

75% of the girls were faster than 50% of the boys.
The range for the girls is smaller.
The median time for the girls is less than for the boys.
The interquartile range for the girls is less than for the boys.
The upper quartile for the girls is less than for the boys.
The lower quartile for the girls is the same as for the boys.

Answers

Tree diagrams

1 a

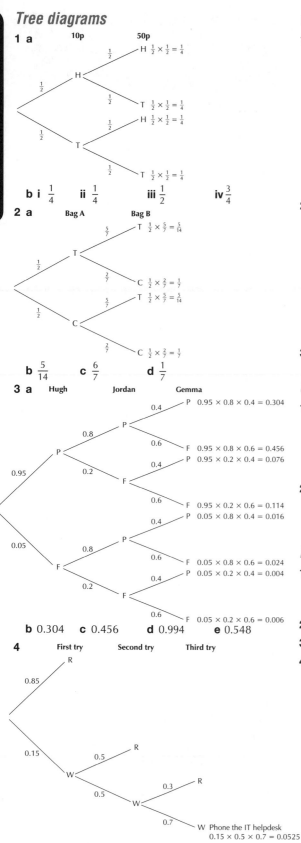

b i $\frac{1}{4}$ **ii** $\frac{1}{4}$ **iii** $\frac{1}{2}$ **iv** $\frac{3}{4}$

2 a

b $\frac{5}{14}$ **c** $\frac{6}{7}$ **d** $\frac{1}{7}$

3 a

b 0.304   **c** 0.456   **d** 0.994   **e** 0.548

4

Probability that she does not need to phone the IT helpdesk: 0.9475

Histograms

1 a

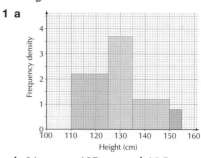

b 84   **c** 127 cm   **d** 16.5 cm   **e** 127.025 cm

2 a

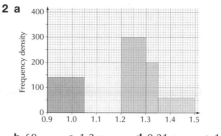

b 69   **c** 1.2 m   **d** 0.21 m   **e** 1.178 m

3 a 9   **b** 3 or 4

Stratified samples

1 a Method **iii** would give a random sample, as all students are equally likely to be chosen.

b Reasons: Method **i** restricts the choice to Year 8 students. Method **ii** restricts the choice to those students who are early for school.

2 25 from Year 7, 26 from Year 8, 29 from Year 9, 30 from Year 10 and 31 from Year 11, giving a total of 141

Probability: combined events

1 a $\frac{1}{13}$ **b** $\frac{12}{13}$

c i $\frac{1}{169}$ **ii** $\frac{144}{169}$ **iii** $\frac{25}{169}$

2 a $\frac{9}{50}$ **b** $\frac{9}{50}$ **c** $\frac{1}{100}$ **d** $\frac{49}{100}$ **e** $\frac{21}{25}$

3 a 0.28   **b** 0.006   **c** 0.288   **d** 0.514

4 $\frac{1}{4}$ **5** $\frac{7}{15}$ **6** $\frac{7}{12}$

ANSWERS TO ASSESSING UNDERSTANDING AND PROBLEM SOLVING

1 45 seconds **2** 4.5 cm **3** $\frac{5}{8}x^2$ **4** 70 cm

5 a $61\frac{1}{3}\pi\,\text{cm}^3$ **b** 3.78 cm

6 a When $y = 0$, $10x - 5$ must $= $ zero, so $x = 0.5$

   b If x is $\frac{4}{3}$ then $\sqrt{16 - 9\left(\frac{4}{3}\right)^2} = 0$, but you cannot divide by zero

   c $-\frac{4}{3}$

7 $AC = \sqrt{125}$ CD and $k = 10$

8 $1 + \sqrt{2}$ cm

9 1 litre

10 120 cm^2

11 (10, 5)

12 (18, 15)

13 $155.5 + 16.5 + 16.5 = 188.5 < 189.5$, will not overflow

14 a $3x^2 - 8x + 8 = 0$

   b There will be no solution since the number inside the square root sign is negative (–32).

15 a 448 tiles **b** 62

16 a 4.8 cm **b** 125 cm^2 **c** 27

17 Small = 6.56 inches^2 per £, big = 7.01 inches^2 per £, so big is better value.

18 3000 cm^3

19 $1000 - 300x + 30x^2 - x^3$

20 Radius = 8 cm, volume = $128\pi\,\text{cm}^3$, mass = $256\pi\,\text{g}$

21 60°

22 Side of square: long part = x, short part = y

   Area of square = $(x + y)^2 = x^2 + 2xy + y^2$

   Using Pythagoras' theorem: $x^2 = \frac{a^2}{2}$, $x = \frac{a}{\sqrt{2}}$

   $y^2 = \frac{b^2}{2}$, $y = \frac{b}{\sqrt{2}}$

   Area of square = $\frac{a^2}{2} + 2 \times \frac{a}{\sqrt{2}} \times \frac{b}{\sqrt{2}} + \frac{b^2}{2}$

   $= \frac{a^2}{2} + ab + \frac{b^2}{2}$

   $= \frac{1}{2}(a^2 + 2ab + b^2)$

   $= \frac{1}{2}(a + b)^2$

23 a $0.1^0 \times 0.9^8 = 0.430...$

   b $1 - (0.1^0 \times 0.9^8 + 0.1^1 \times 0.9^7) = 0.522$

   c $0.1^0 \times 0.9^8 + 0.1^1 \times 0.9^7 \times 0.1^0 \times 0.9^8 = 0.451$

ANSWERS TO SPOT THE ERRORS

1 a $\frac{5}{125} \times 100 = 4$ pence

   b i $\frac{43}{125} \times 40\,300 = \pounds13\,863.20$

   ii $\frac{40\,305}{1.215} = 33\,173$ litres

2 $175 \times 10\,000 = 5 \times 45 \times 40 \times 140F$

   So $1\,750\,000 = 1\,260\,000F$

   $F = 1.3888...$

   The filter type should be internal, over-sized or bigger.

3 If true, then:

   $(2n + 2)^2 - (2n)^2 = 2 \times (2n + 2 + 2n)$

   $4n^2 + 4n + 4n + 4 - 4n^2 = 4n + 4 + 4n$

   $8n + 4 = 8n + 4$ (QED)

4 $\angle ACB = 180° - 128° = 52°$ (angles on a straight line)

   Obtuse $\angle AOB = 2 \times ACB = 104°$ (angle at the centre is double the angle at the circumference when subtended from the same arc)

   Reflex $\angle AOB = 360° - 104° = 256°$ (angles around a point sum to 360°)

5 $p \propto r^2$

   $p = kr^2$

   $7.50 = k \times 1.5^2$

   $k = 3\frac{1}{3}$

   $p = 3\frac{1}{3}r^2$

   $p = 3\frac{1}{3} \times 3^2$

   $p = \pounds30$

6 $2n \begin{array}{l} \nearrow\ n = A \\ \searrow\ 4n = B \end{array}$

   $B - A = 4n - n = 3n$

   Therefore the number is always a multiple of 3.

7 $a^2 + 8b = 121$

   $a^2 - 16b = 1$

   $24b = 120$

   $b = 5$

   $a^2 + 8 \times 5 = 121$

   $a^2 + 40 = 121$

   $a^2 = 81$

   $a = 9$

   So $\sqrt{a^2 - b} = \sqrt{9^2 - 9 \times 5} = \sqrt{36} = 6$

Answers

8 $x^2 + y^2 = (x + 1)^2$

$x^2 + y^2 = x^2 + 2x + 1$

$y^2 = 2x + 1$

$2x + 1$ will be odd for any integer.

Only odd × odd gives an odd number, so y must be odd.

9 a $x = \dfrac{-b \pm \sqrt{b^2 - 4ac}}{2a}$

$b = -8$

$2a = 6$, so $a = 3$

$4ac = 96$

$4 \times 3 \times c = 96$, so $c = 8$

The quadratic equation is: $3x^2 - 8x + 8 = 0$

b Cannot find the square root of a negative number

10 $(4 + \sqrt{5}\,)^2 = (4 + \sqrt{5}\,)(4 + \sqrt{5}\,)$

$= 16 + 4\sqrt{5} + 4\sqrt{5} + 5$

$= 21 + 8\sqrt{5}$

11 LHS $= \dfrac{\sqrt{20} + 10}{\sqrt{5}} \times \dfrac{\sqrt{5}}{\sqrt{5}}$

$= \dfrac{\sqrt{20}\,\sqrt{5} + 10\sqrt{5}}{\sqrt{5}\,\sqrt{5}}$

$= \dfrac{\sqrt{100} + 10\sqrt{5}}{5}$

$= \dfrac{10 + 10\sqrt{5}}{5}$

$= \dfrac{10(1 + \sqrt{5}\,)}{5}$

$= 2(1 + \sqrt{5}\,)$

$=$ RHS (Proved)

12

Score, s	Frequency	Class width	Frequency density
$0 \leqslant s < 20$	10	20	0.5
$20 \leqslant s < 40$	18	20	0.9
$40 \leqslant s < 50$	25	10	2.5
$50 \leqslant s < 60$	20	10	2
$60 \leqslant s < 80$	10	20	0.5
$80 \leqslant s < 100$	5	20	0.25

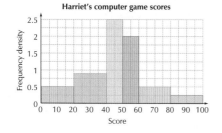

Harriet's computer game scores

13

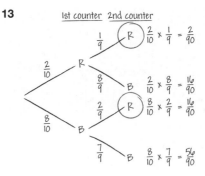

Probability that the second ball is red

$\dfrac{2}{10} \times \dfrac{1}{9} + \dfrac{8}{10} \times \dfrac{2}{9} = \dfrac{2}{90} + \dfrac{16}{90} = \dfrac{18}{90} = \dfrac{1}{5}$

How to interpret the language of exams

If the test says …	The test means …
Calculate	Use a calculator or formal method, e.g. column method.
Deduce	Use more thought than when asked to 'Write down'.
Describe fully	In transformations: reflection – mirror line; translations – vector; rotations – centre, angle and direction; enlargement – scale factor and centre.
Do an accurate drawing	Use a compass to draw lengths; a protractor to measure angles (and a sharp pencil).
Do not use trial and improvement	Use an algebraic method; any sign of trial and improvement will be penalised.
Estimate	*Do not* round numbers, e.g. when asked to find an average speed or the mean average.
Estimate the value of; Find an approximate answer to …	*Do not* work out the exact answer. Round numbers to an easier number, but don't over-round numbers, e.g. 9.8×73 ($10 \times 70 = 700$ is correct; $10 \times 100 = 1000$ is incorrect)
Expand	Rewrite without using brackets (the opposite of 'Factorise', i.e. rewrite with brackets), e.g. Expand $3(x + 6)$ $\qquad 3(x + 6) = 3x + 18$
Expand and simplify	Rewrite without using brackets. Then collect terms. For example: Expand and simplify $2(y + 5) + 4(2y - 3)$ $\qquad\qquad 2(y + 5) + 4(2y - 3) = 2y + 10 + 8y - 12$ $\qquad\qquad\qquad\qquad\qquad = 10y - 2$
Explain; Comment; Give a reason for your answer	Use words or mathematical symbols to explain an answer. A correct answer with no explanation will get zero marks. Examiners may want a keyword or fact, e.g. Susan's triangle has internal angles of 30°, 80° and 80°. Explain why this is wrong. (The internal angles of a triangle add up to 180°. Or: $30 + 80 + 80$ is not equal to 180.)
Explain your answer; You **must** show your working	You will be penalised if you do not show your working.
Express, in terms of …	Use given information to write an expression using only the letter(s) given, e.g. x in terms of z in the equation $12x = z$ $\qquad x = \frac{z}{12}$
Factorise	Put into brackets (opposite of 'Expand'), e.g. Factorise $3x + 9x^2$ $\qquad\qquad 9x^2 + 6x = 3x(3x + 2)$
Factorise fully	Clue, to do more than one factorisation, e.g. find common factor and factorise quadratic.
Give a counter example	Give a numerical or geometrical example to disprove a statement; pick numbers that show a statement is untrue, e.g. Susan says a number is always smaller than its square. Counter example to show she is wrong: $\frac{1}{2}^2 = \frac{1}{2} \times \frac{1}{2} = \frac{1}{4}$
Give an exact value	Give your answer as a square root or a fraction – not as a decimal, which will be recurring, and therefore not exact, e.g. one-third is 0.333333….
Give answer in terms of π* *pi	Equivalent to about 3.14, π has infinite decimals, so a rounded value of π gives an inaccurate answer. Leaving the answer 'in terms of π', gives, in effect, an exact answer, e.g. Use the formula $A = \pi r^2$ to find the area of a circle with radius 5 cm. Answer, in terms of π: $A = \pi r^2$ $\qquad\qquad A = \pi \times 5^2$ $\qquad\qquad A = \pi \times 25$ $\qquad\qquad A = 25\pi$

How to interpret the language of exams

If the test says ...	The test means ...
Give answer to 2 dp* *decimal places	Give your answer to the required accuracy. You will lose marks if you do not do so.
Give answer to a sensible degree of accuracy	Provide an answer that is no more accurate than the values in the question, e.g. if the question has values to 2 sf (significant figures), give the answer to 2 sf or 1 sf. Trigonometrical answers are accepted to 3 sf, as is taught.
Hence	Use the previous answer to proceed.
Hence, or otherwise	Use the previous answer, but if you cannot see how to do so, use another method.
Make x the subject	Rearrange a formula and get x (or other letter) on one side of the equals sign and everything else (including numbers) on the other.
Measure	Use a ruler or protractor to measure a length or angle.
Multiply out	Multiply out using distributive law. (Similar to 'Expand')
Multiply out and simplify	Multiply out using distributive law and then collect terms. (Similar to 'Expand')
Not drawn accurately	Printed next to diagrams to discourage measuring.
Not to scale	Printed next to diagrams (often circles) to discourage measuring.
Prove	Provide a rigid algebraic or geometric proof. (Similar to 'Show that') You are given equivalents, but must show why they are equivalent, e.g. Prove $(x + 3)^2 = x^2 + 6x + 9$ $(x + 3)^2 = (x + 3)(x + 3)$ $(x + 3)^2 = x^2 + 3x + 3x + 9$ $(x + 3)^2 = x^2 + 6x + 9$
Show that	Use words, numbers or algebra to show an answer, e.g. you are given a statement and must explain why it is so.
Simplify	Collect terms or cancel a fraction, e.g. Simplify $7b + 5b$ (added together = $12b$).
Simplify fully	Collect terms and factorise the answer or cancel terms, so an extra numerical or algebraic step is needed. For full marks, the answer must be in its simplest form, e.g. $\frac{8}{16} = \frac{4}{8} = \frac{2}{4} = \frac{1}{2}$
Solve	Find the value(s) of x (or other letter) that makes the equation true and write out the answer, step by step (for full marks), e.g. Solve $3x + 2 = 14$ ($3 \times 4 + 2 = 12 + 2 = 14$)
Use a ruler and compasses	Use a ruler, a straight edge and compasses, e.g. in constructions and loci problems.
Use an algebraic method	Do not use trial and improvement. Working is expected.
Use the graph	Don't calculate; read from the graph, adding lines to show how you got the answer.
Work out	Usually means that a calculation is required, which you may be able to do mentally.
Write down	Answer is clear and does not need any working; it should be in front of you.

Formulae sheet

Volume of a prism = area of cross-section × length

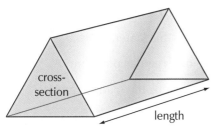

Area of a trapezium = $\frac{1}{2}(a + b)h$

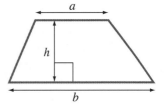

Volume of a sphere = $\frac{4}{3}\pi r^3$

The quadratic equation

The solutions of $ax^2 + bx + c = 0$

Where $a \neq 0$, are given by: $x = \dfrac{-b \pm \sqrt{b^2 - 4ac}}{2a}$

Surface area of a sphere = $4\pi r^2$

In any triangle ABC

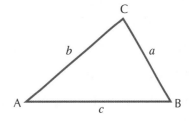

Volume of a cone = $\frac{1}{3}\pi r^2 h$

Sine rule $\dfrac{a}{\sin A} = \dfrac{b}{\sin B} = \dfrac{c}{\sin C}$

Curved surface area of a cone = πrl

Cosine rule $a^2 = b^2 + c^2 - 2bc \cos A$

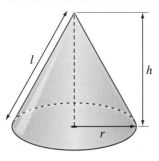

Area of a triangle = $\frac{1}{2}ab \sin C$

Further hints

Area of a circle:

pi times radius squared, or $A = \pi r^2$

Box plots:
often complement the cumulative frequency (cf) curve and show the distribution of data. The two 'whiskers' represent the smallest and largest values and the box is defined by the upper quartile and the lower quartile. The median value is also represented on a box plot.

Circumference of a circle:

pi times diameter, or $C = \pi d$

Cumulative frequency (cf):
Think 'running total' when calculating the cumulative frequency column. To draw a cumulative frequency curve, we plot the cumulative frequency against the end points of the class widths (*not* the midpoints!).

Diameter of a circle:

twice times *pi* times radius, or $D = 2\pi r$

Exterior angle of a regular polygon:

360 degrees divided by number of sides, or $E = \dfrac{360°}{n}$

Factorise:
extracting 'common elements', e.g.

$4x + 6 = 2(2x + 3)$
$10x^2 + 15x = 5x(2x + 3)$
$x^2 + 5x + 6 = (x + 3)(x + 2)$

You can use the 'add and multiply rule' when the coefficient of x^2 is 1. Find two numbers that **add** up to give the coefficient of x and **multiply** to give the constant term. When the coefficient of x^2 is not 1, you can use intuition or the method of factorising in pairs.

Histograms with unequal intervals:
We calculate frequency density as frequency divided by class width to 'dampen down' the wide bars and 'big up' the thin bars. By representing frequency density on the vertical axis, it allows you to ensure that the areas of the bars now represent the true frequency.

Hypotenuse squared:

opposite squared + adjacent squared

Hypothesis:
a testable statement

Interior angle of a regular polygon:

180 degrees minus the exterior angle, or $I = 180° - E$

Leading question:
leads an interviewee in a particular direction

Laws of indices:

$A^m \times A^n = A^{m+n}$
$A^m \div A^n = A^{m-n}$

$A^{-m} = \dfrac{1}{A^m}$

$A^{\frac{m}{n}} = n\sqrt{A^m}$ (e.g. $8^{\frac{2}{3}} = \sqrt[3]{8^2} = \sqrt[3]{64} = 4$)

$A^0 = 1$
$A^1 = A$

Laws of surds:

$\sqrt{A} \times \sqrt{B} = \sqrt{AB}$

$\sqrt{A} \div \sqrt{B} = \sqrt{(A \div B)}$

To simplify a surd you need to find the highest square number factor:

$\sqrt{8} = \sqrt{(4 \times 2)} = \sqrt{4} \times \sqrt{2} = 2\sqrt{2}$

Median (relating to cf):
is halfway up the cumulative frequency curve. The median is also called the second quartile or Q2.

Q1 is the lower quartile, and is one-quarter of the way up the cf curve.

Q3 is the upper quartile, and is three-quarters of the way up the cf curve.

Values for Q1, Q2, and Q3 are found by interpolation (i.e. drawing lines across to the cf curve, hit the curve and read off the values along the horizontal axis).

One linear – one quadratic simultaneous equation: Use this strategy.

   i Substitute from the linear equation into the quadratic equation. Use algebraic techniques to solve $f(x) = 0$.

   ii Note that this often involves solving a quadratic equation.

Probability: P(E) = number of successes or number of outcomes

   For independent events:
   P(A and B) = P(A) × P(B)

   For exclusive events:
   P(A or B) = P(A) + P(B)

Tree diagrams are very helpful for representing probability situations involving more than one event.

The probability values on the branches leading off each node must add up to 1.

Pythagoras: Given two sides on a right-angled triangle, you need to find the length of the third side.

Simultaneous equations: Use this strategy:

   i Can I eliminate one of the variables if I add or subtract the two equations?

   ii If the answer is 'no' equalise the coefficients for one of the variables by multiplying through by a number.

   iii When one variable is found substitute to find the other variable.

Sum of the angles in any polygon:

   $S = 180(n - 2)$ where n = number of sides

The equation of a straight line: $y = mx + c$

   m is the gradient or steepness and c is the y intercept.

If a line is sloping upwards from left to right it has a positive gradient.

If a line is sloping downwards from left to right it has a negative gradient.

If two lines are perpendicular then the product of their gradients is –1.

The interquartile range (IQR): gives the measure of spread of 50% of the data around the average. By comparing the IQR from two samples you can see if one sample is more consistent or has greater variation compared to another sample.

Transformation geometry: You can transform a shape using a reflection, translation, rotation, enlargement or stretch. You must be as specific as possible when describing transformations.

   Reflection: line of reflection must be stated.

   Translation: translation vector must be given.

   Rotation: centre of rotation, number of degrees and direction (clockwise or anti-clockwise) must be given.

   Enlargement: centre of enlargement must be given. Remember that if the scale factor is less than one then the enlargement is, in fact, a reduction.

Trigonometry: involves sides and angles. Work systematically through the 'formula, substitution, work out, check' strategy. Use 'SOH CAH TOA' to find the required angle or side.

Trigonometry involving non-right-angled triangles:

If you have two sides and the included angle, use the cosine rule (provided on the formulae sheet).

If you don't have two sides and the included angle, use the sine rule.

To find an angle given three sides in a non-right angled triangle, rearrange the cosine rule.

To find the area of a triangle given two sides and included angle use:

$$A = \frac{1}{2}ab \sin C$$

Collins

NEW GCSE MATHS

More practice + Smooth progression = Better results

Collins provides you with extra practice and differentiation at <u>all</u> levels.

Remember that in your Higher exam, about 50% of your marks come from questions at the lower grades. Collins has more practice at every grade:

Workbooks
focus on grades G, F, E

Grade C Booster Workbooks
focus on grades D, C, B

Grade A/A* Booster Workbooks
focus on grades B, A, A*

Revision Guides
focus on your Higher or Foundation
exam revision

Revision Apps
revise anywhere with videos, tutorials
and tests

Available on the App
store in October 2010

To order any of these titles, contact us by phone **0844 576 8126**, by fax **0844 576 8131**
or by email: **education@harpercollins.co.uk**

For further information about Collins New GCSE Maths visit our website:
www.collinseducation.com/newgcsemaths